Lecture Notes in Control and Information Sciences

Edited by M. Thoma and A. Wyner

Vol. 62: Analysis and Optimization of Systems
Proceedings of the Sixth International Conference on Analysis and Optimization of Systems
Nice, June 19–22, 1984
Edited by A. Bensoussan, J. L. Lions
XIX, 591 pages. 1984.

Vol. 63: Analysis and Optimization of Systems
Proceedings of the Sixth International Conference on Analysis and Optimization of Systems
Nice, June 19–22, 1984
Edited by A. Bensoussan, J. L. Lions
XIX, 700 pages. 1984.

Vol. 64: Arunabha Bagchi
Stackelberg Differential Games in Economic Models
VIII, 203 pages, 1984

Vol. 65: Yaakov Yavin
Numerical Studies in Nonlinear Filtering
VIII, 273 pages, 1985.

Vol. 66: Systems and Optimization
Proceedings of the Twente Workshop
Enschede, The Netherlands, April 16–18, 1984
Edited by A. Bagchi, H. Th. Jongen
X, 206 pages, 1985.

Vol. 67: Real Time Control of Large Scale Systems
Proceedings of the First European Workshop
University of Patras, Greece, Juli 9–12, 1984
Edited by G. Schmidt, M. Singh, A. Titli, S. Tzafestas
XI, 650 pages, 1985.

Vol. 68: T. Kaczorek
Two-Dimensional Linear Systems
IX, 397 pages, 1985.

Vol. 69: Stochastic Differential Systems – Filtering and Control
Proceedings of the IFIP-WG 7/1 Working Conference
Marseille-Luminy, France, March 12-17, 1984
Edited by M. Metivier, E. Pardoux
X, 310 pages, 1985.

Vol. 70: Uncertainty and Control
Proceedings of a DFVLR International Colloquium
Bonn, Germany, March, 1985
Edited by J. Ackermann
IV, 236 pages, 1985.

Vol. 71: N. Baba
New Topics in Learning Automata Theory and Applications
VII, 231 pages, 1985.

Vol. 72: A. Isidori
Nonlinear Control Systems: An Introduction
VI, 297 pages, 1985.

Vol. 73: J. Zarzycki
Nonlinear Prediction Ladder-Filters for Higher-Order Stochastic Sequences
V, 132 pages, 1985.

Vol. 74: K. Ichikawa
Control System Design based on Exact Model Matching Techniques
VII, 129 pages, 1985.

Vol. 75: Distributed Parameter Systems
Proceedings of the 2nd International Conference, Vorau, Austria 1984
Edited by F. Kappel, K. Kunisch, W. Schappacher
VIII, 460 pages, 1985.

Vol. 76: Stochastic Programming
Edited by F. Archetti, G. Di Pillo, M. Lucertini
V, 285 pages, 1986.

Vol. 77: Detection of Abrupt Changes in Signals and Dynamical Systems
Edited by M. Basseville, A. Benveniste
X, 373 pages, 1986.

Vol. 78: Stochastic Differential Systems
Proceedings of the 3rd Bad Honnef Conference, June 3–7, 1985
Edited by N. Christopeit, K. Helmes, M. Kohlmann
V, 372 pages, 1986.

Vol. 79: Signal Processing for Control
Edited by K. Godfrey, P. Jones
XVIII, 413 pages, 1986.

Vol. 80: Artificial Intelligence and Man-Machine Systems
Edited by H. Winter
IV, 211 pages, 1986.

Lecture Notes in Control and Information Sciences

Edited by M. Thoma and A. Wyner

131

S. M. Joshi

Control of Large Flexible Space Structures

Springer-Verlag
Berlin Heidelberg GmbH

Author
Suresh M. Joshi
510 Blue Ridge Hunt Rd.
Hampton, VA 23666
USA

ISBN 978-3-540-51467-1 ISBN 978-3-540-48143-0 (eBook)
DOI 10.1007/978-3-540-48143-0

Originally published by Springer-Verlag Berlin Heidelberg New York in 1989

2161/3020-543210 Printed on acid-free paper.

TO

Shyamala,

to

my parents,

and to

my parents-in-law

The views and contents of this book are solely those of the author and not of the National Aeronautics and Space Administration.

Preface

In the history of science and technology, theoretical advancements and practical applications have always had a mutually stimulating effect. The aerospace sciences area is no exception to this. As new, more advanced aerospace systems emerge, the requirements for better performance, higher reliability and less cost, are becoming more stringent, requiring further theoretical developments in many different disciplines. One such area, which is becoming increasingly important, is the design, construction and operation of very large satellites in Earth orbit. This upcoming class of satellites, which would require large structures to be deployed, assembled, and maintained in space, is expected to have a profound impact on the quality of life here on Earth, by making quantum improvements in communications, astronomy, and Earth observation; by establishing permanent human presence in space; and by facilitating manned exploration of the solar system.

In order to achieve the required performance, it is of utmost importance to be able to control such structures in space with high precision in attitude and shape. Control of such structures is a challenging problem because of their special dynamic characteristics, which result from their large size and light weight.

This book is intended to be a research monograph that presents the problems encountered in controlling large, flexible space structures (LFSS), and some of the control synthesis methods developed by the author, as well as results of their application to realistic spacecraft models. In the interest of economy of space, we have not included all the exhaustive research done in this area by several researchers, but rather have concentrated on some of our own results. However, we have attempted to provide an adequate bibliography for those wishing to refer to works of other researchers. It is hoped that the book will stimulate research activity in control theory with application to flexible spacecraft, and that it will be also be useful to practitioners working in the area of satellite control. The book will also serve as an introductory text for persons intending to start working in the area of flexible spacecraft control.

The organization of the book is as follows: The basic mathematical models of LFSS are presented in the Chapter 1, and the difficulties and challenges involved in designing control laws are explained. We concentrate on using finite-dimensional models [i.e., involving ordinary differential equations (ODE)], although infinite- dimensional models [i.e., described by partial differential equations (PDE)] are briefly discussed and their relationship to ODE models is pointed out. The reason for following the ODE approach is that models for most realistic spacecraft can only be generated using finite- element computer programs such as NASTRAN, and are in the ODE form.

Chapters 2 and 3 address the problem of precision attitude control system synthesis for LFSS. In most applications it is essential to control the attitude of the LFSS to specified precision in order for the LFSS to function as required. For instance, a communications antenna must be pointed with extremely high accuracy towards the target on Earth. It must also retain accurate shape of its reflector, and accurate relative position between the feeds and the reflector. The attitude control problem we consider thus includes the control of not only pointing, but also shape distortions resulting from flexibility. Two approaches are considered for attitude control system design. The first approach (presented in Chapter 2) is the "dissipative" controller, which utilizes a number of actuators and sensors at same (or close) locations, distributed throughout the structure. Using the dissipativity type properties of LFSS, we prove that such controllers are robust to modeling errors as well as actuator and sensor nonlinearities and phase lags. We also present results of application of such controllers to a large, flexible space antenna. A particular type of actuator, the Annular Momentum Control Device (AMCD), which has the inherent characteristic of actuator/sensor collocation, is described, and its stability and robustness properties are analyzed.

Chapter 3 considers the use of linear-quadratic-Gaussian (LQG)- type controllers for attitude control of LFSS. The time-domain design approach is considered first, and matrix norm bounds on modeling errors, which ensure stability, are obtained. Multivariable frequency domain techniques for designing controllers which are robust to modeling errors are next considered. In particular, the LQG/loop transfer recovery (LTR) method is modified for application to LFSS, and results for a large space antenna model are presented. The stability of LQG-type controllers in the presence of realistic actuator nonlinearities is investigated, and expressions for the regions of attraction and ultimate boundedness are obtained.

Chapter 4 addresses some related topics, namely, parameter identifiability studies, and the large attitude-angle maneuvering problem. Areas for future research are also discussed.

To demonstrate the controller synthesis methods, we have attempted to make use of numerical examples based on realistic LFSS. In particular, the LFSS model we have used most often is that of a large space antenna, namely, the 122 m diameter "hoop/column" antenna. The antenna model serves as a thread which is common to all the chapters, and hopefully simplifies the explanation of various methods discussed.

I would like to thank my wife Shyamala for her patient support and encouragement. I would also like to thank NASA Langley Research Center for granting me permission to write this book. Many thanks are also due to Ms. Barbara Jeffrey for her expert typing of the final manuscript.

Suresh M. Joshi

Hampton, Virginia

December 1988

Contents

Chapter 1

Introduction

Future utilization of space is expected to require large space structures in low-Earth as well as geosynchronous orbits. Examples of such future missions include: electronic mail systems, Earth observation systems, mobile satellite communication systems, solar power satellites, large optical reflectors, and space stations. Such missions typically require large antennas, platforms and solar arrays. These missions would be feasible because of the launching capability of the Space Shuttle, which can be expanded by augmentation with Orbit Transfer Vehicles (OTV) for placement in the geosynchronous orbit.

The dimensions of such structures would typically range from 50 meters (m) to possibly several kilometers (km). For example, one mobile personal communication system concept, for the entire continental United States, would require a space antenna with a diameter of 122 m. To establish such structures in space at minimum cost would require that their weight be minimized. It will also be necessary to compactly package them in sub-assemblies, each of which is deployable and can fit in the Shuttle cargo bay. Some of the structures (e.g., space station) will require on-orbit assembly using components such as deployable beams.

Because of their light weight and expansive sizes, these structures will tend to have extremely low-frequency, lightly damped structural (elastic) modes. Natural frequencies of the elastic modes would be generally closely spaced, and some natural frequencies may be lower than the controller bandwidth. In addition, the elastic mode parameters (natural frequencies, mode shapes and damping ratios) would not be known accurately.

For these reasons, control systems design for large flexible space structures (LFSS) is a difficult and challenging problem.

Two of the most important control problems for LFSS are: i) Fine-pointing of LFSS in space with the required precision in attitude (that is, the three pointing angles) and shape, and ii) Large angle maneuvering ("slewing") the LFSS to orient to a different target. The performance requirements for both of these problems are usually very high. For example, in some applications, it is necessary to maneuver the LFSS quickly to acquire a new target on the Earth. This has to be done in the minimum possible time, and with the minimum fuel expenditure, while keeping the elastic motion and accompanying stresses within acceptable limits. In geosynchronous applications, the anticipated Earth-pointing slew would roughly translate into a 20 degree maneuver in 10 seconds or less. (20 degrees would cover the entire diameter of the Earth at the geosynchronous altitude). The slew angles are much larger for low-Earth orbits. Once the target is acquired, the LFSS must point to it with the required precision. For example, for the mobile communication system mission, the antenna will have to be pointed to within 0.03 degree root mean square (RMS). The requirements for other missions vary, but some are expected to be even more stringent-on the order of 0.01 arc-second. This book mainly addresses the problem of fine-pointing.

The basic problems in pointing control of LFSS have been known for several years in the context of control of conventional spacecraft, which are relatively rigid, but which have sufficient flexibility to necessitate consideration in the design process. Examples of the early studies in this area include Saturn I and IB [Kie.64], Saturn V Apollo [Pin.69], and Skylab [Har.76]. [Gev.70] also includes an analysis of the problems arising from lightly damped elastic modes. However, the distinguishing characteristic of LFSS is their highly prominent structural flexibility, which makes LFSS a new class of spacecraft. (A detailed literature survey on dynamics and control of LFSS may be found in [Nur.84]).

In this chapter, the basic mathematical models of LFSS are presented. They include infinite-dimensional models described by partial differential equations (PDE's), and finite dimensional approximations, i.e., involving ordinary differential equations (ODE's). The relationship between PDE and ODE models is briefly discussed, although we concentrate on using only ODE models in the remainder of the book. The reason for following the ODE approach is that most engineering models are generated using finite element computer programs such as NASTRAN, and are in the ODE form. Two examples of finite element models are presented, and the procedure for using finite element data to generate state space models is explained. Controllability and observability of ODE models are discussed, and the difficulties and challenges involved in control systems synthesis are explained.

1.1. Mathematical Models of Large Flexible Space Structures

Two types of dynamic models have been been used in the literature for LFSS. The first type consists of continuous models, which result in distributed-parameter (infinite-dimensional) systems represented by partial differential equations with appropriate boundary conditions. The second type of models consist of finite- dimensional systems, represented by ordinary differential equations.

1.1.1 Infinite-Dimensional Models

A class of LFSS can be generically described by a system of partial differential equations (PDEs) such as [Bal.82]:

$$m(s)\frac{\partial^2}{\partial t^2}u(s,t) + D\frac{\partial}{\partial t}u(s,t) + Au(s,t) = F(s,t) \tag{1}$$

where $u(s,t)$ is a displacement (translational or rotational) of the structure from its equilibrium position, as a function of space variable s and time t. $m(s)$ is the mass density, A is a time-invariant differential operator, whose domain consists of all smooth functions satisfying (1) with appropriate boundary conditions, and is thus dense in the infinite dimensional Hilbert space $L_2(\Omega)$, where Ω denotes the structure. The operator A is generally self-adjoint and non-negative. $F(s,t)$ is the distribution of the applied generalized force (i.e., forces and moments). D represents the inherent damping operator, which is a property of the structure (materials, joint design, etc).

In most cases, the characterization of the damping operator is not straightforward. The standard method is to assume "proportional damping", which is not well-defined even for simple structures like a uniform beam. Modeling of the damping term is treated in [Hug.82], [Bal.86], [Bal.87], and is still an area of active research. Because the damping is usually very small, and because a reliable damping model is not available, D is usually assumed to be zero (a null operator). In cases where finite-dimensional approximations are used, proportional (or viscous) damping is added later, after such a model has been obtained assuming zero damping.

In most cases, the operator A has a discrete spectrum, and its eigenvalue equation can be written as:

$$A\phi_k(s) = \omega_k^2 m(s)\phi_k(s) \quad k = 1, 2, 3, ... \tag{2}$$

where $\phi_k(s)$ and ω_k^2 are the eigenvector and the eigenvalue. ω_k is called the natural (or structural) frequency, and $\phi_k(s)$ is called the "mode shape" function of the kth mode.

In most real applications, the applied forces and moments are produced by localized or "point devices", that is, they are applied at discrete points on the structure. Assuming that forces $f_i(t)$ are applied at points $s_i (i = 1, 2, .., m)$, the force distribution can be expressed as:

$$F(s,t) = \sum_{i=1}^{m} f_i(t)\delta(s - s_i) \tag{3}$$

where $\delta(.)$ denotes the Dirac delta function. A torque T at a point r can be approximated as two equal and opposite forces, F and $-F$, applied at: $s = r - \varepsilon$, and $s = r + \varepsilon$ respectively, where $F = T/2\varepsilon$. These forces have to be included in the appropriate translation component of the force distribution $F(s,t)$. As $\varepsilon \to 0$, the forces approach the doublet: $T\partial\delta(s - r)/\partial s$.

In this formulation we have omitted the gravitational field, orbital effects, atmosphereic drag, solar pressure, etc. These can be treated as external disturbances which are usually predictable and repeatable, or have small magnitudes, and they and are not considered herein. Examples of applied forces and moments include reaction control jets, reaction wheels, momentum wheels, control moment gyros (CMG) etc.

The sensors aboard an LFSS may include star sensors or Sun sensors (attitude measurement), rate gyros (attitude rate measurements), accelerometers, laser/optical ranging devices (to measure relative deflection), etc. They are usually "point" devices, and the measurements (translational or rotational displacements) at location s_j can be expressed as:

$$y_j(t) = u(s_j, t) \tag{4}$$

The rate or velocity measurements are merely the time-derivatives of the measurements in (4).

Using the fact that A is self-adjoint, it can be shown that the eigenvectors ϕ_k are orthogonal in the sense that

$$< \phi_i, A\phi_j >= \omega_j^2 < \phi_i, m\phi_j >= 0 \quad \text{for} \quad i \neq j \tag{5}$$

where $< .,. >$ represents the usual $L_2(\Omega)$-inner product. If ϕ_i are normalized by dividing each ϕ_i by: $< \phi_i, m\phi_i >$, we have

$$< \phi_i, m\phi_j >= \delta_{ij} \tag{6}$$

where δ_{ij} represents the Kronecker delta function. This infinite-dimensional model can be transformed into a "modal" model consisting of infinite number of second-order ODE's. To illustrate this let us consider a uniform free-free beam whose planar motion (i.e., bending along an axis pependicular to the beam longidunial axis) is described by the Euler-Bernoulli PDE:

$$\bar{m}\frac{\partial^2 u}{\partial t^2} + EI\frac{\partial^4 u}{\partial s^4} = \sum_{i=1}^{m} f_i(t)\delta(s - s_i) + \sum_{j=1}^{p} T_j(t)\frac{\partial}{\partial s}\delta(s - r_j) \tag{7}$$

where $\bar{m}$ and EI denote the mass per unit length and the bending stiffness of the beam, and u is the transverse displacement. The boundary conditions for both ends to be completely free are: $\partial^2 u(s,t)/\partial s^2 = \partial^3 u(s,t)/\partial s^3 = 0$ at $s = 0$ and $s = L$, where L is the length of the beam. (Such boundary conditions, which involve second or third order spatial derivatives of u, are called "dynamic" or "natural" boundary conditions because they are obtained by applying force or moment balance on the boundary. Boundary conditions involving spatial derivatives of order 0 or 1 are called "geometric" boundary conditions; e.g., if the beam is clamped at one end, $u(s,t) = \partial u(s,t)/\partial s = 0$ at $s = 0$). Suppose eigenvectors $\phi_k(s)$ are normalized so that

$$< \phi_k, \bar{m}\phi_k >= 1 \tag{8}$$

they form an orthonormal basis of the Hilbert space $L_2(\Omega)$. Thus $u(s,t)$ has a unique representation:

$$u(s,t) = \sum_{k=1}^{\infty} q_k(t)\phi_k(s) \tag{9}$$

where $q_k(t)$ is called the "modal amplitude" for the kth mode. Substituting for $u(s,t)$ from (9) into (7) and taking inner product of both sides of (7) with $\phi_i(s)$ yields [using (2)]:

$$\ddot{q}_i + \omega_i^2 q_i = \sum_{j=1}^{m} \phi_i(s_j) f_j(t) + \sum_{j=1}^{p} \phi_i'(r_j) T_j(t) \quad i = 1,2,3,.... \tag{10}$$

where a prime denotes the spatial derivative ("slope"). In writing (10) we have used (8) and the fact that ϕ_k are orthogonal. We have also made use of the fact that the inner proudct of ϕ_i with the doublet term $\partial\delta(s - r_j)/\partial s$ in (7) yields: $\phi_i'(r_j)T_j(t)$.

The displacement at location s_j is given by:

$$u(s_j,t) = \sum_{k=1}^{\infty} \phi_k(s_j) q_k(t) \tag{11}$$

and the angular displacement at s_j is given by:

$$u'(s_j,t) = \sum_{k=1}^{\infty} \phi_k'(s_j) q_k(t) \tag{12}$$

The translational and rotational velocities are simply the time- derivatives of these terms (Eqs. 11 and 12), which would involve the $\dot{q}_k$, the modal velocities.

In the above discussion we have illustrated the transformation of an infinite-dimensional system into an infinite set of ODE's for a simple free-free beam. For general three-dimensional structures, however, mere formulation of the problem in the infinite-dimensional setting is quite formidable. (For example, see [Bal.85] wherein a realistic LFSS model is formulated as an abstract wave equation). In theory, the procedure used above for the simple beam can also be used for more complex spacecraft ([Rob.85], [Jos.84]). However, the eigenvalue problem is generally very difficult to solve for complex non-symmetric structures, and it is usually necessary to use approximate methods, which result in a finite number of ODE's.

The operator equation (1) describes the linear motion of the LFSS, which consists of elastic and rigid-body (zero frequency) modes. The rigid-body modes include three translations and three rotations (with respect to three orthogonal axes). These modes are very important because the primary objective is to control the basic rigid-body attitude (three pointing angles) and position of the spacecraft. In many cases the rigid motion is too large to be in the linear range, although the elastic motion is small. In some such cases an approximate model may be obtained by super-imposing the small elastic motion on the nonlinear rigid motion. For the precision attitude control problem, both rigid and elastic motion are small (about the zero equilibrium position) and mutually uncoupled. It is usually simpler to write the rigid and elastic models separately and to combine them. The linearized rigid-body equations are:

$$M\ddot{z} = \sum_{i=1}^{m} f_i \tag{13}$$

$$J_s\ddot{\alpha} = \sum_{i=1}^{m} R_i \times f_i + \sum_{j=1}^{p} T_j \tag{14}$$

where z denotes the 3×1 vector representing the position of the center of mass (c.m.) in an inertial coordinate system, M_s and J_s denote the mass and the 3 x 3 moment of inertia matrix (about the c.m.), $\alpha = (\phi, \theta, \psi)^T$ represents the (rigid-body) attitude angles about X, Y and Z axes, and R_i denotes the 3×1 coordinate vector of the point of application of force f_i with respect to the c.m ($a \times b$ denote the cross-product of 3-vectors a and b). The elastic ODE model is given by (10), (wherein only elastic modes are included). Thus the complete linearized ODE representation is given by Eqs. (10), (13) and (14).

The sensors mentioned previously are physical devices that measure the total motion, which includes the rigid and elastic motion. The inertial position of a point on the LFSS is:

$$y_j(t) = z(t) - d_j \times \alpha(t) + \sum_{k=1}^{\infty} \phi(d_j) q_k(t) \tag{15}$$

where d_j is the position vector of the point, and ϕ_k is the (3×1) translational mode shape for the kth mode. The total attitude at that point (which is measured, for example, by a star sensor) is given by:

$$\beta_j = \alpha(t) + \sum_{k=1}^{\infty} \psi_k(d_j) q_k(t) \tag{16}$$

where ψ_k denotes the 3×1 rotational mode shape for the kth mode. An attitude rate sensor at point d_j would measure the time-derivative of (16).

In precision attitude control problems, the rigid-body translation is usually not important. It occurs only when force actuation devices such as reaction jets are used for attitude control and orbit correction. Application of torques or moments is a better way of controlling the attitude, since it leaves the c.m. position, and therefore the orbital parameters (altitude, velocity, etc.) unaffected. In order to study the attitude control problem, we only need to consider (10) and (14), and the sensor equations.

1.1.2 Approximate Methods for Finite-Dimensional Models

LFSS designed for real missions are expected to be quite complex and would lead to eigenvalue problems for which closed-form solutions are not possible. There are a number of approximate methods which can be used for such cases. The methods are essentially schemes for discretization of continuous systems, and can be divided into two classes. The first class represents the solution as a finite summation of functions of space variables, multiplied by generalized coordinates, which are functions of time. The second class of methods uses a lumped-mass approximation of continuous systems. (See [Mei.80] or other standard texts on vibrational analysis for a detailed discussion of approximate methods). Herein we briefly discuss only the first class of methods, which is used more widely. In particular, the best known methods in the first class are: the Rayleigh-Ritz method, the method of weighted residuals, and the finite element method.

Rayleigh-Ritz Method

In this method, which is applicable only to self-adjoint systems, an approximate solution (eigenvector) of the eigenvalue problem is expressed as;

$$u(s) = \sum_{i=1}^{n} v_i u_i(s) \tag{17}$$

where v_i are real coefficient, and u_i are known smooth, independent functions which satisfy the geometric boundary conditions, but not necessarily the differential equation of the system.

Suppose ϕ and λ are an eigenvector and the corresponding eigenvalue, which satisfy the operator eigenvalue equaiton:

$$A\phi = m\lambda\phi \tag{18}$$

Taking inner product with ϕ ,

$$< \phi, A\phi >= \lambda < \phi, m\phi >= \lambda < \sqrt{m}\phi, \sqrt{m}\phi > \tag{19}$$

or

$$\lambda = \frac{< \phi, A\phi >}{< \sqrt{m}\phi, \sqrt{m}\phi >} \tag{20}$$

Since u in (17) is an approximation and not an eigenvector, (20) will not hold if ϕ is replaced by u. Define the Rayleigh quotient as:

$$R(\bar{v}) = \frac{\langle \sum_1^n v_i u_i, \sum_1^n v_i A u_i \rangle}{\langle \sqrt{m} \sum_1^n v_i u_i, \sqrt{m} \sum_i^n v_i u_i \rangle} \tag{21}$$

where $\bar{v} = (v_1, v_2, ..., v_n)^T$. It can be shown that $\bar{v}$ which gives the best approximation for the eigenvector is obtained by making $\partial R / \partial \bar{v} = 0$, which yields:

$$(K - \lambda M)\bar{v} = 0 \tag{22}$$

where $K = K^T \geq 0$ and $M = M^T > 0$ are n × n matrices, which are called the "stiffness matrix" and the "mass matrix" respectively. Eq. (22) represents an algebraic (rather than operator) eigenvalue problem, for which the n eigenvalues λ_i and the corresponding eigenvectors $\bar{v}_i, (i = 1, 2, ..., n)$, can be readily computed. The approximate structural frequencies are then given by $\omega_i = \sqrt{\lambda_i}$, and the approximate eigenfunctions are given by using the coefficient vectors $\bar{v}_i$ in (17).

The "assumed-modes" method is closely related to the Rayleigh-Ritz method, and may be considered to be a variation of the latter. In this method, the solution is assumed to be of the form:

$$u(s,t) = \sum_{i=1}^{n} \psi_i(s) \eta_i(t) \tag{23}$$

where ψ_i are the known admissible functions (i.e., satisfying the geometric boundary conditions) which represent the best guess of the mode-shape functions, and η_i are the generalized coordinates, which are functions of time, rather than constants as in (17). The subsequent steps in the assumed-modes method include writing the expressions for the kinetic and potential energy as functions of ψ's and η's, and application of the Lagrangian formulation, which results an equation of the following form:

$$M\ddot{\eta} + K\eta = 0 \tag{24}$$

where $M = M^T > 0$ and $K = K^T \geq 0$. The M and K matrices obtained in this manner are identical to those in (22). Furthermore, this set of differential equations can be recast [Mei.80] as an algebraic eigenvalue problem of the form of Eq.(22). For the eigenvalue problem in (22), suppose ϕ_i and $\omega_i^2 (i = 1, 2, ..., n)$ represent the eigenvectors and the eigenvalues. The eigenvectors can be shown to be M-orthogonal, that is,

$$\phi_i^T M \phi_j = 0 \text{ if } i \neq j$$

Suppose ϕ_i are normalized so that $\phi_i^T M \phi_i = 1$. Denoting

$$\Phi = [\phi_1, \phi_2, ..., \phi_n], \text{ and } \Lambda = \text{diag}[\omega_1^2, \omega_2^2, ..., \omega_n^2]$$

it can be shown that

$$\Phi^T M \Phi = I \quad \text{and} \quad \Phi^T K \Phi = \Lambda \tag{25}$$

Using the transformation:

$$\eta(t) = \Phi q(t) \tag{26}$$

where $q = (q_1, q_2, ..., q_n)^T$, Eq. (24) is transformed into the following set of uncoupled second-order ODE's:

$$\ddot{q} + \Lambda q = 0 \tag{27}$$

$q_i(t)$ and $\phi_i(s), (i = 1, 2, ..., n)$ are called natural modal amplitudes and mode shapes of the system, respectively.

Method of Weighted Residuals

This method is more general than the Raleigh-Ritz method because its application is not limited to self-adjoint systems. In this method, an approximate solution is assumed as in (17), and the "residual" of the eigenvalue equation is defined as:

$$R(u,s) = Au(s) - \lambda m(s)u(s) \tag{28}$$

Let $\psi_i (i = 1, 2, ..., n)$ represent an independent set of functions in $L_2(\Omega)$. For a given n, we attempt to obtain weighting functions $\psi_i (i = 1, 2, ..., n)$, all of which are orthogonal to R. As $n \to \infty$, R will be orthogonal to the entire space $L_2(\Omega)$, which implies: $R \to 0$. By imposing the condition that all ψ_i are orthogonal to R, one can obtain the algebraic eigenvalue problem of the form (22), wherein M and K are generally not symmetric. Depending on the choice of $\{\psi_i\}$, a variety of methods for solution can be obtained. The best known of these methods is perhaps the Galerkin method, in which $\psi_i = u_i (i = 1, 2, ..., n)$. If A is self-adjoint, the resulting algebraic eigenvalue problem is identical to that obtained by the Rayleigh-Ritz method.

Finite Element Method

The finite element method is by far the most commonly used method for modeling LFSS, because of its versatality in handling highly complex structures, and its amenability to automation by digital computer. The basic philosophy of this method is to divide a continuous suystem into a number of elements using fictitious dividing lines. The points of intersection of the dividing lines are referred to as "nodes" or "joints". Each joint has a certain number of degrees of freedom (DOF), up to a maximum of six DOF (three translational and three rotational). For example, for planar motion of a slender beam element, the displacement (bending) perpendicular to the axis, is expressed as a function of the space variable s as:

$$w(s,t) = \sum_{i=1}^{4} \phi_i(s) u_i(t) \tag{29}$$

where $u_1(t)$ and $u_2(t)$ are the translational and rotational displacements at one joint (one end) of the element, and $u_3(t)$ and $u_4(t)$ are those at the other end. $\phi_i(s)$ are the shape functions which are usually obtained using the differential equation governing static beam bending. In contrast to the previous methods, the coefficients $u_i(t)$ in (29) are the actual physical coordinates, which is a nice feature of the finite element method. Lagrangian formulation is subsequently applied wherein the joint forces are determined as functions of the applied force distribution and forces produced by the adjacent elements. The problem finally reduces to:

$$M\ddot{u} + Ku = 0 \tag{30}$$

where $u(t)$ is of dimension n, given by

$$n = \sum_{i=1}^{n_j} n_{DOF}(i) \tag{31}$$

n_j being the number of joints, and $n_{DOF}(i)$ is the number of DOF for the ith joint. Eq. (30) has the same form as (24). For given types of elements (e.g., beam, plate, shell, membrane, etc.), the standard shape functions are known; therefore, dividing an LFSS into such elements, and using the interconnection information, makes the generation of K and M highly amenable to machine computation. Subsequent solution of the eigenvalue problem using standard numerical techniques is also straightforward. If point-forces and torques are applied at each joint, the final "modal" model has the form (after using the transformation: $u(t) = \Phi q(t)$ as in (26)):

$$\ddot{q}(t) + \Lambda q(t) = \Gamma^T f(t) \tag{32}$$

$$y(t) = \Gamma q(t) \tag{33}$$

where q is the $n \times 1$ modal amplitude vector, Γ is the $n \times n$ generalized mode-shape matrix, f is the vector of generalized applied forces (consisting of forces and torques), and y is the $n \times 1$ displacement vector, consisting of translational and rotational displacements at each joint.

The finite element method gives both the rigid and the elastic modes. That is, $q(t)$ in Eq. (32) also includes zero-frequency modes. However, it is usually more convenient to use only the elastic-mode portion of the finite element model, and to augment it with the rigid-body equations [Eqs. (13) and (14)] obtained independently. It is also more accurate to do this because zero-frequency modes are usually computed by the finite element method as some combinations of the basic translational and rotational modes, and with small non-zero frequencies due to numerical errors. Thus we assume from here on that $q(t)$ in (32) contains only elastic modal amplitudes.

Models obtained in this manner cannot predict the inherent structural damping. It is customary to add damping terms to Eq. (32), i.e., after the eigenvalue problem has been solved. The most commonly used damping term is proportional (or viscous) damping, which modifies (32) as follows:

$$\ddot{q}(t) + D\dot{q}(t) + \Lambda q(t) = \Gamma^T f(t) \tag{34}$$

where

$$D = 2\,\mathrm{diag}\,(\rho_1\omega_1, \rho_2\omega_2, ..., \rho_n\omega_n) \tag{35}$$

where ρ_i and ω_i denote the inherent damping ratio and the natural frequency of the ith elastic mode. For LFSS, ρ_i's are typically on the order of 0.001-0.01. The complete model is then given by (13), (14) and (34), and can be easily represented in the state variable form.

Examples of Finite-Element Models

Example 1: Large, thin, completely free, flat plate - Consider a 100 ft $\times$ 100 ft $\times$ 0.1 in., free-free-free-free (i.e., free at all the boundaries) aluminum plate. Suppose it is divided into 24 $\times$ 24 equal square plate elements. Each corner of each element represents a "joint"; thus there are 625 joints. Suppose the plate is rigid about the axis perpendicular to its plane. Then the elastic motion consists of only the out-of-plane elastic displacement and two transverse rotations. That is, each joint has three degrees of freedom ($n_{DOF} = 3$ for all joints). The dimension of the eigenvalue problem is: $n = 625 \times 3 = 1875$. However, we usually need to compute fewer eigenvalues (frequencies) and eigenvectors (generalized mode-shapes). Table 1 shows the first 44 natural frequencies of the plate, computed using the SPAR computer package [Whe.78]. Thus the Γ matrix is 1875 $\times$ 44. The ith mode shape can be plotted by giving q_i a reasonable value (while making all other q_j's zero), and computing the out-of-plane displacements at each joint.

Table 1. Elastic mode frequencies for large, thin, flat plate

MODE	FREQ(RAD/SEC)
1	0.05499
2	0.08002
3	0.09911
4	0.14211
5	0.14211
6	0.24948
7	0.24948
8	0.26008
9	0.28286
10	0.31515
11	0.43068
12	0.43068
13	0.47824
14	0.50003
15	0.53689
16	0.53689
17	0.62422
18	0.65958
19	0.68808
20	0.80973
21	0.80973
22	0.83371
23	0.87377
24	0.87982
25	0.87982
26	0.99216
27	0.99216
28	1.1483
29	1.1922
30	1.1996
31	1.2194
32	1.2250
33	1.2532
34	1.2532
35	1.3742
36	1.4082
37	1.4871
38	1.4871
39	1.6059
40	1.6059
41	1.6972
42	1.6972
43	1.7111
44	1.7523

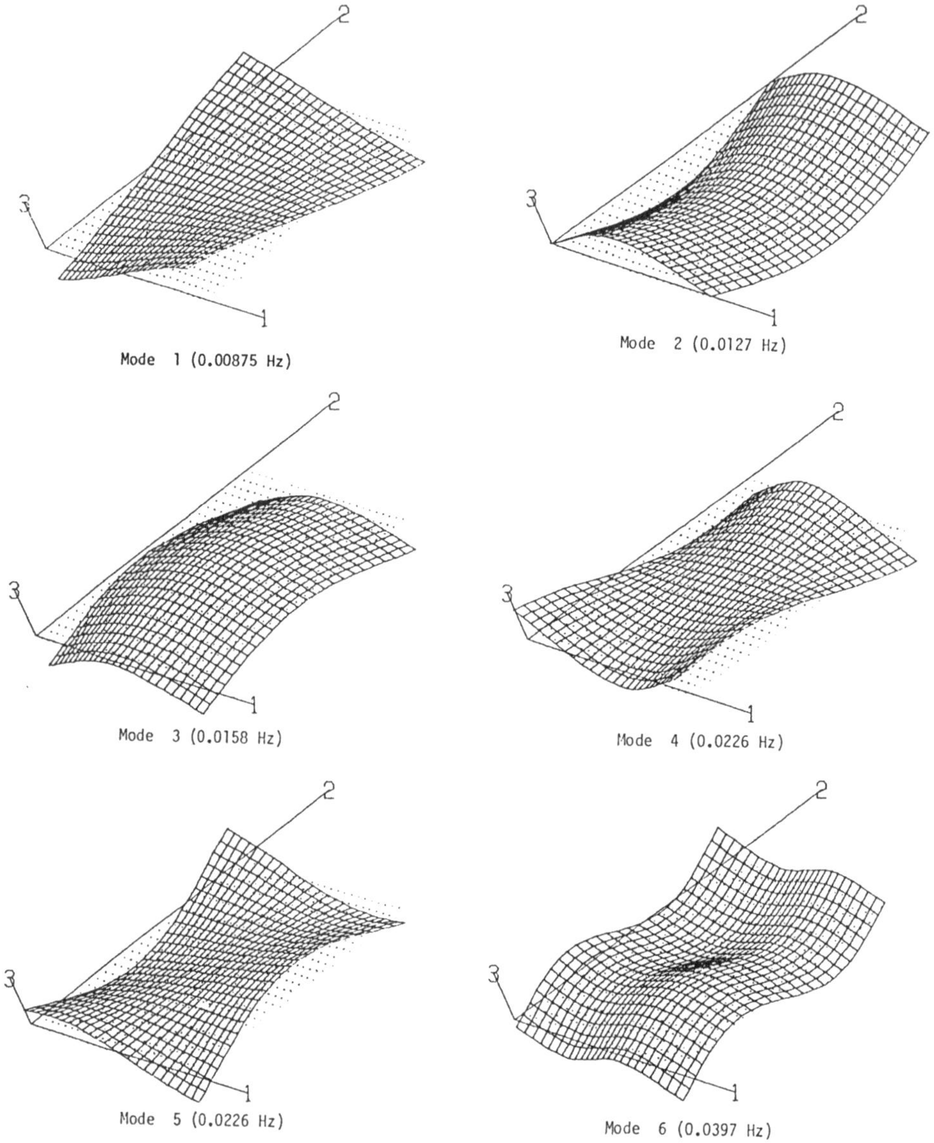

Figure 1. Mode-shape plots for large, thin, flat plate

STRUCTURAL MODE NO. 23, FREQ = .8377E+00 RAD/SEC (.1396E+00 HZ)

JOINT NO.	COORDINATES FT.	PHI-Z IN/IN	THETA-X RAD/IN	THETA-Y RAD/IN
501	(16.67, 0.00)	-.3199E-01	.2460E-02	-.1103E-02
502	(16.67, 4.17)	.1028E+00	.2811E-02	-.2013E-03
503	(16.67, 8.33)	.2389E+00	.2494E-02	-.4764E-03
504	(16.67, 12.50)	.3411E+00	.1479E-02	-.2013E-03
505	(16.67, 16.67)	.3799E+00	.1160E-04	-.1109E-04
506	(16.67, 20.83)	.3428E+00	-.1479E-02	.6243E-04
507	(16.67, 25.00)	.2406E+00	-.2528E-02	.2271E-04
508	(16.67, 29.17)	.1044E+00	-.2811E-02	-.9027E-04
509	(16.67, 33.33)	-.2473E-01	-.2258E-02	-.2142E-03
510	(16.67, 37.50)	-.1092E+00	-.1082E-02	-.2874E-03
511	(16.67, 41.67)	-.1278E+00	.2912E-03	-.2714E-03
512	(16.67, 45.83)	-.8315E-01	.1374E-02	-.1641E-03
513	(16.67, 50.00)	.1239E-05	.1783E-02	.1281E-07
514	(16.67, 54.17)	.8316E-01	.1374E-02	.1641E-03
515	(16.67, 58.33)	.1278E+00	.2912E-03	.2714E-03
516	(16.67, 62.50)	.1092E+00	-.1082E-02	.2875E-03
517	(16.67, 66.77)	.2473E-01	-.2258E-02	.2142E-03
518	(16.67, 70.83)	-.1044E+00	-.2811E-02	.9029E-04
519	(16.67, 75.00)	-.2406E+00	-.2528E-02	-.2269E-04
520	(16.67, 79.17)	-.3428E+00	-.1479E-02	-.6241E-04
521	(16.67, 83.33)	-.3980E+00	.1109E-04	.1111E-04
522	(16.67, 87.50)	-.3411E+00	.1479E-02	.2013E-03
523	(16.67, 91.67)	-.2389E+00	.2494E-02	.4764E-03
524	(16.67, 95.83)	-.1028E+00	.2811E-02	.7875E-03
525	((16.67,100.0)	.3200E-01	.2460E-02	.1103E-02

Figure 2. Typical finite element model data

The plots of the first six mode-shapes are shown in Figure 1. Figure 2 shows a typical output data page for the finite-element model, which can be used to obtain the values of elements of the generalized mode-shape matrix. For example, the row corresponding to joint number 501 gives its coordinates, followed by $\Gamma_{(500\times3+1),23}$, $\Gamma_{(500\times2+1),23}$, and $\Gamma_{(500\times3+3),23}$. The complete model, which consists of all the $1875\times44(=82,500)$ entries of Γ, would occupy about 1200 pages similar to Figure 2. However, we normally need the elements of Γ only corresponding to the sensor and actuator locations, which are far fewer than 82,500. If there are m actuators, the Γ matrix in (34) is replaced by an $m \times n_q$ matrix Γ_f (assuming only n_q modes are included in the model). If there are ℓ sensors, the sensor output is given by (33) wherein Γ is replaced by the $\ell \times n_q$ matrix Γ_s. If $\ell = m$, and if the actuators and sensors are collocated and compatible (i.e., force actuator and position sensor, or torque actuator and attitude sensor), then $\Gamma_s = \Gamma_f$.

Many finite-element computer outputs customarily express the data in inch-pound-second (in.-lb.-sec.) units. That is, forces, torques, displacements and rotations are expressed in: lb., in.-lb., in., and radians respectively. If these units are used, Γ_s and Γ_f are not equal for the collocated actuator/sensor case. Conversion to fundamental units (ft.-lb.-sec.) can be accomplished by [Jos.80c] i) dividing the rows of Γ corresponding to displacement (translations) by $\sqrt{12}$, and ii) multiplying the rows of Γ corresponding to rotations by $\sqrt{12}$. After this modification, $\Gamma_s = \Gamma_f$ for the collocated case.

Example 2: Large space antenna- Consider the 122 m diameter hoop/column antenna concept ([Rus.80], [Sul.82]), shown schematically in Figure 3. The antenna concept consists of a deployable central mast attached to a deployable hoop by cables held in tension. A secondary drawing surface is used to produce the desired contour of the radio- frequency (RF) reflective mesh. The shaping of the RF surface is accomplished by mesh shaping ties. The deployable mast consists of a number of telescoping sections which are deployed by means of a cable drive system. The hoop consists of 48 rigid segments, and is deployed by four motor drive units. The reflective mesh, which is made of knit gold-plated molybdenum wire, is attached to the hoop by quartz or graphite fibers. Precision shaping of the RF mesh (e.g., spherical, parabolic, etc.) is accomplished by control cords aattached to the mesh through the secondary drawing surface.

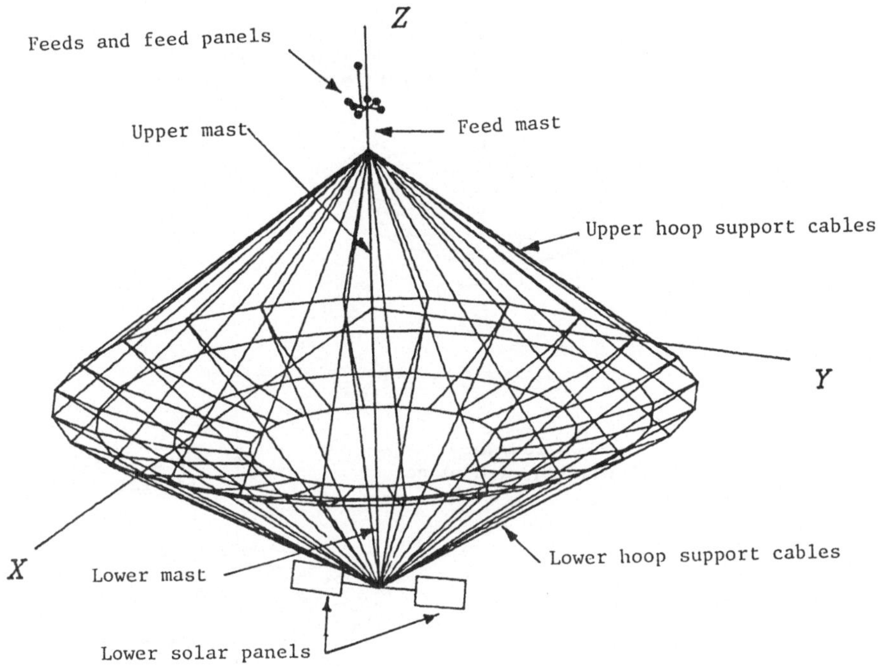

Figure 3. Hoop/column antenna schematic

Table 2. Hoop/column antenna parameters

Mass= 4544.3 Kg.

Inertia about axes through center of mass (Kg-m^2)

$I_{xx} = 5.724 \times 10^6$ $I_{yy} = 5.747 \times 10^6$

$I_{zz} = 4.383 \times 10^6$ $I_{xz} = 3.906 \times 10^4$

$I_{xy} = I_{yz} = 0$

Mode no.	1	2	3	4	5	6	7	8	9	10
Freq. rad/sec	0.75	1.35	1.70	3.18	4.53	5.59	5.78	6.84	7.4	8.78

Mode no.	11	12	13	14	15	16	17	18	19	20
Freq. rad/sec	10.85	11.24	15.05	15.4	15.75	15.85	16.04	18.84	18.84	18.99

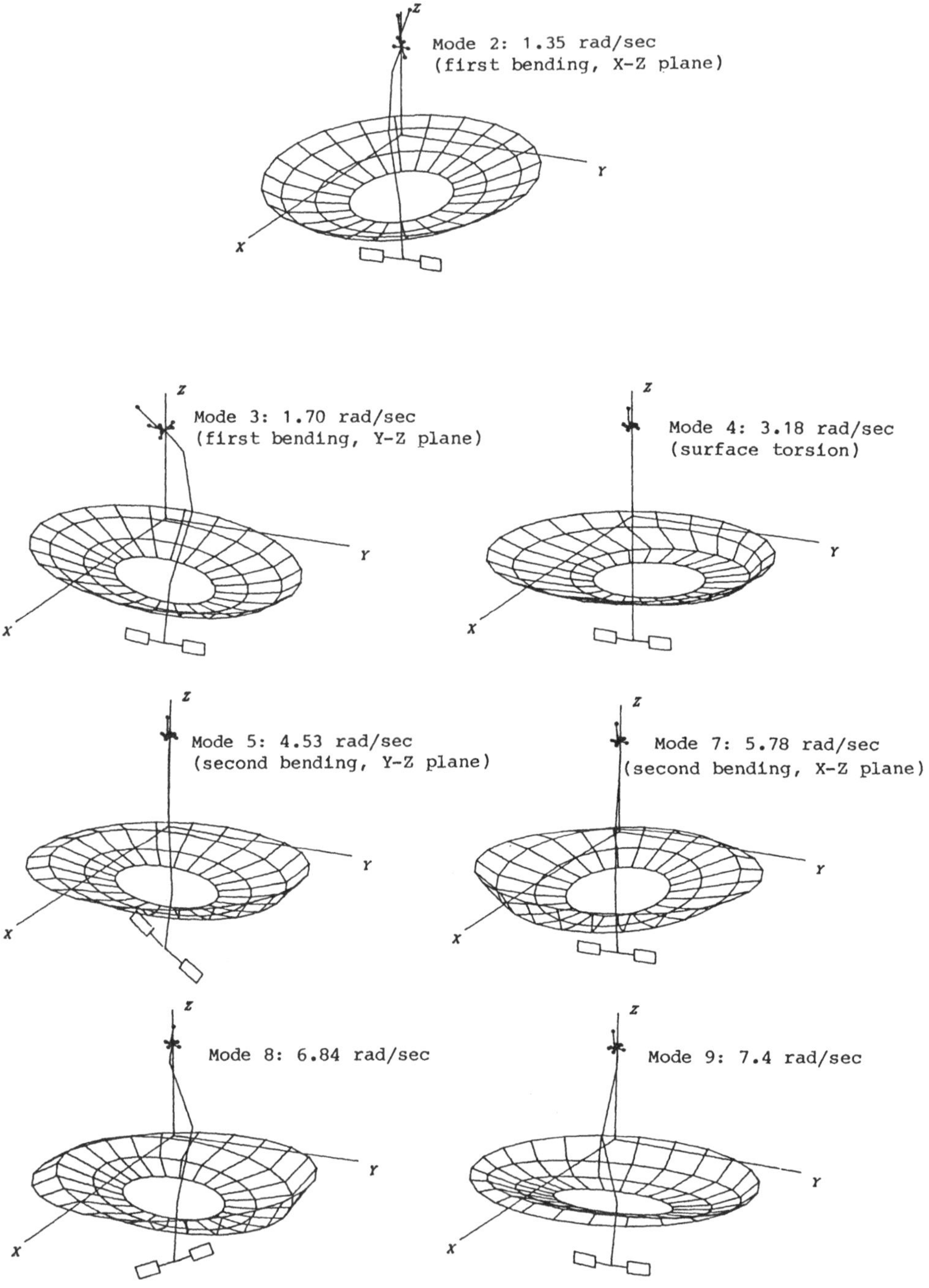

Figure 4. Typical antenna mode shapes

A 20-elastic-mode finite-element model of the antenna was obtained [Sul.82] and the resulting elastic mode frequencies are shown in Table 2, along with other parameters of the antenna. Using the mode-shape data for 112 joints (each with DOF = 6) on the mast, the feeds, the feed panels, the solar panels, and the RF mesh, a few typical mode-shapes are plotted in Figure 4. The format for the finite-element model data is very similar to that for the plate (Figure 2), but each joint has six degrees of freedom, compared to three DOF for each joint for the plate.

1.1.3 Controllability and Observability of Finite-Dimensional Models

A finite dimensional model of an LFSS can be expressed in the state-space form as:

$$\dot{x} = Ax + Bu \tag{36}$$

where

$$x = (q_{rb}^T, \dot{q}_{rb}^T, q_1, \dot{q}_1, q_2, \dot{q}_2, ..., q_{n_q}, \dot{q}_{nq})^T \tag{37}$$

u is the $m \times 1$ control vector consisting of applied forces and torques, q_{rb} is the 6×1 vector of rigid translations and rotations, and q_i denotes the ith modal amplitude.

$$A = \text{diag}(A_{rb}, A_1, A_2, ..., A_{nq}) \tag{38}$$

$$A_{rb} = \begin{bmatrix} 0_6 & I_6 \\ 0_6 & 0_6 \end{bmatrix} \tag{39}$$

(0_k and I_k denote the $k \times k$ null and identity matrices respectively).

$$A_i = \begin{bmatrix} 0 & 1 \\ -\omega_i^2 & -2\rho_i\omega_i \end{bmatrix} \tag{40}$$

$$B = \begin{bmatrix} B_{rb} \\ B_1 \\ B_2 \\ \vdots \\ B_{nq} \end{bmatrix} \tag{41}$$

$$B_i = \begin{bmatrix} 0 \\ b_i^T \end{bmatrix} \tag{42}$$

As an example, consider the attitude control problem with only torque actuators, and attitude and rate sensors. In that case, the rigid translational motion is ignorable, and $q_{rb} = (\phi, \theta, \psi)^T$ which are the three rigid-body rotation (Euler) angles, and

$$A_{rb} = \begin{bmatrix} 0_3 & I_3 \\ 0_3 & 0_3 \end{bmatrix} \tag{43}$$

Consider a single 3-axis torque actuator, which yields:

$$B_{rb} = \begin{bmatrix} 0_3 \\ J_s^{-1} \end{bmatrix} \tag{44}$$

where J_s is the 3×3 moment of inertia matrix. If a single 3-axis attitude sensor is used, the sensor output is:

$$y = Cx \tag{45}$$

where

$$C = \begin{bmatrix} C_{rb} & C_1 & C_2 & \dots & C_{nq} \end{bmatrix} \tag{46}$$

$$C_{rb} = \begin{bmatrix} I_3 & 0_3 \end{bmatrix} \tag{47}$$

$$C_i = \begin{bmatrix} c_{i(3\times1)} & 0_{3\times1} \end{bmatrix} \tag{48}$$

If a three-axis attitude rate sensor is used,

$$C_{rb} = \begin{bmatrix} 0 & I_3 \end{bmatrix} \tag{49}$$

$$C_i = \begin{bmatrix} 0 & c_i \end{bmatrix} \tag{50}$$

If both attitude and rate sensors are used (at the same location),

$$C_{rb} = I_6 \tag{51}$$

$$C_i = \text{diag}(c_i, c_i) \tag{52}$$

The following theorem gives conditions for controllability.

Theorem. *The system given by Eq. (36) is controllable if and only if (iff) all of the following conditions are satisfied:*

i) *B_{rb} is of full rank*

ii) *Each b_i [see (42)] corresponding to distinct eigenvalues of A has at least one nonzero entry*

iii) *For each multiple eigenvalue λ_j of multiplicity ν, the matrix*

$$\bar{B}_j = \begin{bmatrix} b_{j1}^T \\ b_{j2}^T \\ \cdot \\ \cdot \\ b_{j\nu}^T \end{bmatrix}$$

is of rank ν, where b_{jk}^T is the row of the B matrix corresponding to the second row of the kth 2×2 block for λ_j .

Proof. The proof can be obtained by straightforward application of the Popov-Belevitch-Hautus (PBH) rank test [Kai.80]. ■

(The symbol " ■ " is used to denote the end of a proof, or, when no proof is given, the end of a theorem statement.)

As an example, for the attitude control problem with only torque actuators, condition (i) would be satisfied iff there is at least one torque actuator per axis. Condition (ii) would be satisfied iff the rotational mode shape (or "mode-slope") for each mode is nonzero at the location of at least one actuator. Condition (iii) needs to be tested only when there are more than one elastic modes with the same structural frequency. This is not an uncommon occurrence in the case of symmetric structures (for example, see Table 1).

Similar necessary and sufficient conditions can be obtained for observability in an entirely analogous manner, and that discussion is omitted. However, it is worth noting for the attitude control problem that the rigid-body modes are not observable using attitude-rate sensors alone (i.e., with no attitude sensor). The system can be observable if at least three attitude sensors (one per axis) are used, even when rate sensors are absent.

1.2 Problems in Controller Design for Large Space Structures

The main problem considered in this book is the attitude (or fine-pointing) control problem, which implies controlling the rigid rotational modes and suppressing the elastic vibration. The objective of the controller is to

1) quickly damp out the pointing errors resulting from step disturbances (such as thermal distortion resulting from entering or leaving Earth's shadow), or nonzero initial conditions (e.g., resulting from the completion of a large-angle attitude maneuver), and
2) maintain the attitude as close as possible to the desired attitude in the presence of noise and disturbances.

The first objective translates into the closed-loop bandwidth requirement, while the second one translates into minimizing the root mean square (RMS) pointing error. In addition, it is required that the elastic motion be very small; i.e., the RMS shape distortions must be below prescribed limits. For application such as the large communications antenna, the typical bandwidth would be 0.1 rad/sec, with at most 4 second time constant for all the elastic modes (closed-loop). Typical allowable RMS errors are: 0.03 deg. pointing error, and 6 mm surface distortion.

The problems encountered in designing an attitude controller are:

1) An adequate model of an LFSS is of high order because it contains a large number of elastic modes. Therefore, a practically implementable controller has to be of

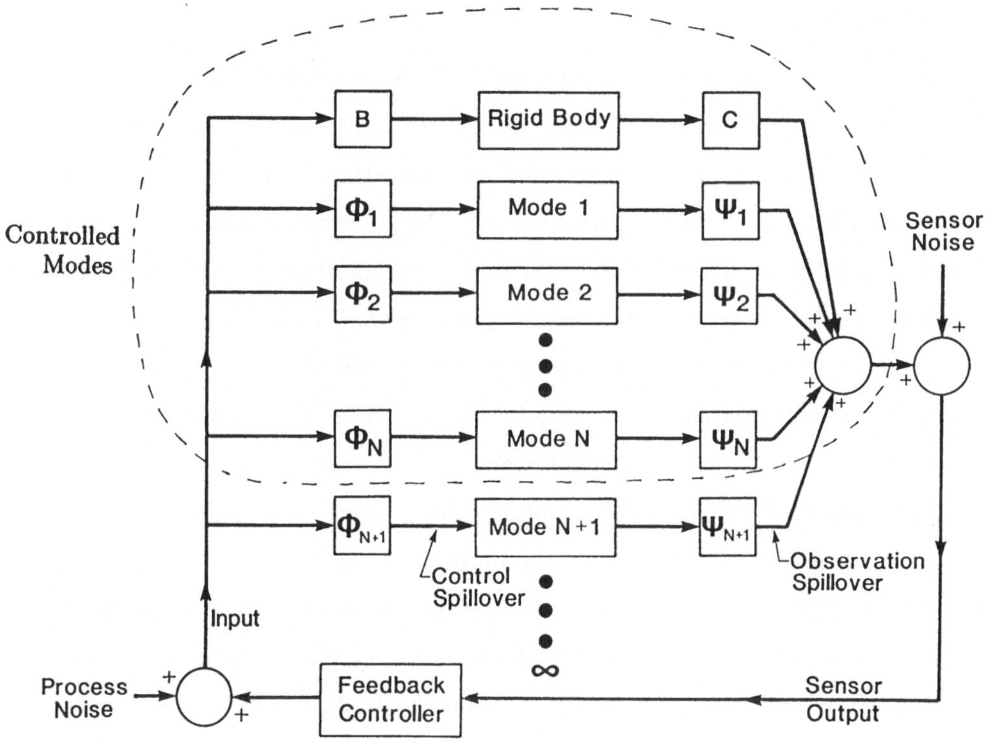

Figure 5. Effect of modal truncation

reduced order.

2) The inherent energy dissipation (damping) is very small.

3) The elastic frequencies are low and closely-spaced.

4) The parameters (frequencies, damping ratios and mode shapes) are not known accurately apriori.

The simplest controller design approach would be to truncate the model at a certain mode, and to use the truncated "design model" to design a controller, which, of course, would be of reduced order. This approach is routinely used for controlling relatively rigid conventional spacecraft, wherein only the rigid modes are retained in the design model. Second-order filters are included in the loop to attenuate the contribution of the elastic modes. This approach does not generally work for LFSS because the elastic modes are much more prominent. Figure 5 shows the effect of using a truncated design model. When building a control loop around the "controlled" modes (which are included in the design model, one also unintentionally builds a feedback loop around the truncated (or "residual") modes. The resulting control system may make the closed-loop system unstable. The inadvertent excitation of the residual modes and the unwanted contribution of the residual modes to the sensed output were termed by Balas [Bal.77] respectively as "control spillover" and "observation spillover". The spillover terms may cause reduction in performance, and even instability leading to catastrophic failure.

In addition to the truncation problem, the designer usually lacks accurate knowledge of the parameters. Approximate methods (such as finite element) are known to give reasonably accurate estimates of the frequencies and mode shapes only for the first few modes, and can provide no estimates of inherent damping ratios. Pre- mission ground-testing for parameter estimation would be infeasible because LFSS are not designed to withstand the gravitational force, and because the test facilities required (e.g., vacuum chamber) would be excessively large. Another consideration in controller design is that the actuators and sensors have nonlinearities and finite response times. In view of these problems, the attitude controller must be a "robust" one, that is, it must maintain at least stability, and perhaps performance, despite modeling errors, uncertainties, nonlinearities and component failures.

Chapter 2

A Class of Robust Dissipative Controllers

The main problem in designing a controller for an LFSS is the modeling error ("spillovers") arising from truncating the model, and from the lack of accurate knowledge of the parameters. In this chapter, we consider a class of controllers which circumvent these problems. Such controllers, termed "dissipative" or "collocated" controllers, consist of compatible pairs of actuators and sensors which may be distributed throughout the LFSS. For example, an attitude and an attitude rate sensor is located at the same point as a torque actuator; displacement and velocity sensors are collocated with a force actuator. We consider two types dissipative controllers: 1) collocated attitude controller (CAC), and 2) controllers using velocity feedback. The second type of controllers includes a) collocated damping enhancement controllers (CDEC), and b) total velocity feedback controllers (TVFC). A CAC is designed to control both the rigid modes and the elastic modes, whereas the function of a CDEC and TVFC is to damp out the elastic motion. The CDEC is used to enhance the structural damping without affecting the rigid modes, while the TVFC additionally stabilizes the rigid motion in the sense that all rigid-body rates also tend to zero. Both types of controllers use output feedback, and therefore are simple to implement.

2.1 Description of Dissipative Controllers

The linearized mathematical model of the system is as follows:

Rigid modes: Center of mass (c.m.) translation:

$$m_s\ddot{\xi} = \sum_{i=1}^{m_f} f_i \tag{1}$$

where m_s is the LFSS mass, ξ is the displacement of the c.m., and f_i $(i = 1, 2, ..., m_f)$ are the applied forces, each being a 3-vector consisting of X, Y and Z components.

Rotation:

$$J_s\ddot{\alpha} = \sum_{i=1}^{m_f} r_i \times f_i + \sum_{j=1}^{m_T} T_j \tag{2}$$

where J_s is the 3×3 moment of inertia matrix, $\alpha = (\phi, \theta, \psi)^T$ is the rigid- body Euler angle vector about X, Y and Z axes, r_i is the 3×1 coordinate vector of the location of the applied force f_i , and $T_j(j = 1, 2, ..., m_T)$ is the 3×1 torque at location j.

Elastic motion [from eq. (34), Chapter 1]:

$$\ddot{q} + D\dot{q} + \Lambda q = \Psi^T F + \Phi^T T \tag{3}$$

where q is the $n_q \times 1$ modal amplitude vector, and Ψ and Φ are (respectively) $3m_f \times n_q$ and $3m_T \times n_q$ mode-shape and mode-slope (i.e., rotational mode shape) matrices. F and T denote the applied force and torque vectors.

We first consider the case where the actuators and sensors are "ideal" (i.e., linear and instantaneous). Eqs. (1)-(3) can be combined as:

$$A_s\ddot{p} + B_s\dot{p} + C_s p = \Gamma^T u \tag{4}$$

where

$$A_s = \begin{bmatrix} m_s & 0 & 0 \\ 0 & J_s & 0 \\ 0 & 0 & I_{n_q} \end{bmatrix} \tag{5}$$

(The symbol I_k is used to denote the $k \times k$ identity matrix).

$$B_s = \text{diag}\left(0_3, \quad 0_3, \quad D_{n_q \times n_q}\right) \tag{6}$$

where D is the $n_q \times n_q$ symmetric matrix representing the inherent structural damping, and 0_k denotes the $k \times k$ null matrix. Since some damping, no matter how small, is always present, we assume $D > 0$.

$$C_s = \text{diag}\left(0_3, \quad 0_3, \quad \Lambda_{n_q \times n_q}\right) \tag{7}$$

where Λ is the diagonal matrix of squared elastic mode frequencies, and

$$\Gamma^T = \begin{bmatrix} I_3, & I_3, & \ldots, & I_3, & 0_3, & 0_3, & \ldots, & 0_3 \\ \tilde{r}_1, & \tilde{r}_2, & \ldots, & \tilde{r}_{m_f}, & I_3, & I_3, & \ldots, & I_3 \\ \Psi_1^T, & \Psi_2^T, & \ldots, & \Psi_{m_f}^T, & \Phi_1^T, & \Phi_2^T, & \ldots, & \Phi_{m_T}^T \end{bmatrix} \tag{8}$$

$\tilde{r}$ denotes the cross product matrix of the 3-vector $r = (r_x, r_y, r_z)^T$

$$\tilde{r} = \begin{bmatrix} 0 & -r_z & r_x \\ r_z & 0 & -r_y \\ -r_x & r_y & 0 \end{bmatrix}$$

$$p = (\xi^T, \quad \alpha^T, \quad q^T)^T \tag{9}$$

$$u = \left(f_1^T, f_2^T, \ldots, f_{m_f}^T, T_1^T, T_2^T, \ldots, T_{m_T}^T\right)^T \tag{10}$$

In (8), Ψ_j and Φ_j denote the $3 \times n_q$ mode-shape and mode- slope matrices for the jth actuator location. Suppose m_f 3-axis translational position and velocity sensors are placed at the same locations as the m_f force actuators, and m_T 3-axis attitude and rate sensors are placed at the same locations as the m_T torque actuators. Then the 3×1 translational displacement vector at the kth sensor location will be:

$$y_{fk} = \xi - r_k \times \alpha + \Psi_k q \tag{11}$$

Similarly, the 3×1 attitude vector at sensor location k will be:

$$y_{Tk} = \alpha + \Phi_k q \tag{12}$$

Denoting the sensor output vector by y_p ,

$$y_p = \left(y_{f_1}^T, y_{f_2}^T, ..., y_{f_{m_f}}^T, y_{T_1}^T, ..., y_{T_{m_T}}^T\right)^T \tag{13}$$

It can be readily verified that

$$y_p = \Gamma p \tag{14}$$

It can be seen that the matrices A_s, B_s, and C_s are symmetric. In addition, A_s is positive definite, and B_s and C_s are positive semidefinite. Note that the "Γ" in (14) is the same "Γ" which appears in (4). It is this property of the system (which is due to collocation of the actuators and sensors), togethter with the symmetry and non-negativity of A_s, B_s, and C_s, that enables the design of a class of robustly stable output feedback contrtollers.

The output of the rate senors is given by:

$$y_r = \Gamma \dot{p} \tag{15}$$

In reality, the sensor output will be contaminated by additive observation noise. However, since the noise is of no consequence in the stability analysis, we ignore it for the moment. Consider the control law:

$$u = -G_p y_p - G_r y_r \tag{16}$$

where G_p and G_r denote the $3(m_f + m_T) \times 3(m_f + m_T)$ proportional and rate gain matrices.

The Collocated damping enhancement controller (CDEC) is similar to the CAC described above, except that its purpose is to enhance structural damping using velocity feedback, without affecting the rigid modes. In the CDEC mode, the rigid components of the translational and rotational velocities are removed from the velocity feedback signal. For example, suppose there are two torque actuators and collocated attitude rate sensors at two distinct locations on the structure. Then the elastic motion can be described by:

$$\ddot{q} + D\dot{q} + \Lambda q = \Phi_1^T T_1 + \Phi_2^T T_2 \tag{17}$$

If we put the constraint that $T_2 = -T_1 = T$, then the right hand side of the above equation becomes: $(\Phi_2 - \Phi_1)^T T$. In addition, if we subtract the outputs of the two rate gyro signals, we get:

$$y = (\Phi_2 - \Phi_1)\dot{q} \tag{18}$$

By following this procedure using paired actuator and sensor locations, the rigid motion can be removed from the equations. The elastic motion can be expressed in the form:

$$\ddot{q} + D\dot{q} + \Lambda q = \tilde{\Gamma}^T u \tag{19}$$

where $\tilde{\Gamma}$ denotes the matrix obtained by subtracting the mode slopes. The modified rate sensor output vector is:

$$\tilde{y}_r = \tilde{\Gamma}\dot{q} \tag{20}$$

It should be noted that the use of force actuators in this manner would generally affect the rigid rotational modes; therefore, torque actuators are more suitable for this purpose.

The CDEC control law is given by:

$$u = -G_r \tilde{y}_r$$

If G_r is diagonal, the resulting CDEC configuration is called "member damper" controller [Can.78].

The total velocity feedback controller (TVFC) is identical to the CAC, except that the position gain $G_p = 0$ in eq. (16). Thus this controller includes the feedback of both the rigid and the elastic components of the rates. The TVFC can accomplish damping enhancement and attitude stabilization (but not control), in the sense that the rigid body rates tend to zero.

2.2 Stability Properties of Dissipative Controllers

In this section we investigate the stability properties of dissipitave controllers with the assumption that the sensors and actuators are perfect, i.e., linear and instantaneous. We first consider the CAC. From (4), (14), (15) and (16), the closed-loop equations become:

$$A_s\ddot{p} + \bar{B}_s\dot{p} + \bar{C}_s p = 0 \tag{21}$$

where

$$\bar{B}_s = B_s + \Gamma^T G_r \Gamma, \quad \text{and} \quad \bar{C}_s = C_s + \Gamma^T G_p \Gamma \tag{22}$$

The following theorem proves the stability of the closed-loop system for the CAC.

Theorem 1. *Suppose G_p and G_r are symmetric, and $G_p > 0$. Then the closed-loop system given by eqs. (21) and (22) is stable in the sense of Lyapunov if $G_r \geq 0$, and is asymptotically stable if $G_r > 0$.*

Proof. We first prove that $\bar{C}_s$ is positive definite. Consider the quadratic form:

$$W(p) = p^T \bar{C}_s p = q^T \Lambda q + p^T \Gamma^T G_p \Gamma p$$

Since $\Lambda > 0$, $W(p)$ can be 0 only if $q = 0$, and $p^T \Gamma^T G_p \Gamma p = 0$; i.e., only if

$$(x_{rb}^T, q^T)\, \Gamma^T G_p \Gamma \begin{bmatrix} x_{rb} \\ q \end{bmatrix} = 0$$

where $x_{rb} = (\xi^T, \alpha^T)^T$. Thus $W(p)$ can be 0 only if

$$x_{rb}^T(\mathcal{I}^T G_p \mathcal{I})x_{rb} = 0, \tag{23}$$

where

$$\mathcal{I} = \begin{bmatrix} I_3, & I_3, & \ldots, & I_3 & 0_3, & 0_3, & \ldots, & 0_3 \\ \tilde{r}_1, & \tilde{r}_2, & \ldots, & \tilde{r}_{m_f}, & I_3, & I_3, & \ldots, & I_3 \end{bmatrix}^T$$

Clearly, since the coefficient matrix in (23) is positive definite, $W(p)$ can be zero only if $x_{rb} = 0$. Thus we conclude that $W(p) = 0$ only if $p = 0$, and that $\bar{C}_s$ is positive definite.

Now consider the Lyapunov function:

$$V(p, \dot{p}) = p^T \bar{C}_s p + \dot{p}^T A_s \dot{p} \tag{24}$$

V is positive definite. Using (21),

$$\begin{aligned} \dot{V} &= 2p^T \bar{C}_s \dot{p} + 2\dot{p}^T A_s \ddot{p} \\ &= -2\dot{p}^T \bar{B}_s \dot{p} \end{aligned} \tag{25}$$

If $G_r \geq 0$, $\dot{V}$ is negative semidefinite. Therefore, the system is stable in the sense of Lyapunov.

Suppose $G_r > 0$. Using (22) in (25),

$$\dot{V} = -2\dot{p}^T \Gamma^T G_r \Gamma \dot{p} - 2\dot{q}^T D \dot{q}$$

Therefore, for $\dot{V} = 0$, it is necessary that $\dot{q} = 0$, and hence that $\dot{x}_{rb} = 0$. Considering the closed-loop equation, this implies

$$\bar{C}_s p = 0 \tag{26}$$

which can happen (since $\bar{C}_s > 0$) only when $p = 0$. Thus $\dot{V} < 0$ along all the trajectories, and the system is asymptotically stable. ■

An important significance of the above result is that the stability is maintained *regardless of the number of modes in the model*, and *regardless of the inaccuracy in the knowledge of the parameters.* Thus the spillover problem is completely avoided, and the parameters do not have to be known to get stability.

In the above analysis we have considered the complete motion of the LFSS, including translation and rotation. In practice, the rotational modes are more important because they determine the attitude, which is the variable to be controlled with high precision. The only consideration for the c.m. translation is the decay of the orbit (i.e., reduction in altitude and velocity) because of atmospheric drag. Any significant decay in orbit, which would require corrective action, happens very slowly, perhaps over a period of days or weeks. The orbit correction is accomplished by periodically (once every few days or weeks) by firing reaction jets for a few seconds or minutes, until the nominal altitude and velocity are re-established. Because of their infrequent nature, orbital corrections are not considered in the attitude control problem during normal operation.

For attitude control, torque actuators are better suited than force actuators for the following reasons:

i) torque actuators do not cause c. m. translation

ii) torque actuators can produce precision torques, whereas reaction jets are basically on-off devices, and can at best produce (using several vernier jets) quantized forces; the quantization error is usually large enough to make high precision attitude control unachievable.

iii) torque actuators use electric power from the solar generators onboard the spacecraft, the supply of which is virtually unlimited, whereas force actuators require stored fuel which has to be resupplied from the Earth.

For these reasons, we shall ignore the translation of the c.m., and consider only the rotational rigid modes and the elastic motion in the LFSS dynamics, in much of our analysis. We shall also limit our attention mostly to torque actuators and attitude and rate sensors. *Without changing the notation*, we write the modified equations of motion as:

$$A_s\ddot{p} + B_s\dot{p} + C_s p = \Gamma^T u \tag{27}$$

where

$$p = (\alpha^T, q^T)^T \tag{28}$$

$$A_s = \text{diag}[J_s, I_{n_q}] \tag{29}$$

$$B_s = \text{diag}[0_3, \quad D_{n_q \times n_q}] \tag{30}$$

$$C_s = \text{diag}[0_3, \quad \Lambda_{n_q \times n_q}] \tag{31}$$

$$\Gamma^T = \begin{bmatrix} I_3 & \cdots\cdots & I_3 \\ \Phi_1^T & \cdots\cdots & \Phi_{m_T}^T \end{bmatrix} \tag{32}$$

$$u = (T_1^T, \quad T_2^T, \quad \ldots, \quad T_{m_T})^T \tag{33}$$

The $m \times 1$ (where $m = 3m_T$) attitude and rate sensor output vectors (ignoring noise) are:

$$y_p = \Gamma p \tag{34}$$

$$y_r = \Gamma \dot{p} \tag{35}$$

With these definitions, we use the same control law (eq. 16); i.e.,

$$u = -[G_p y_p + G_r y_r] \tag{36}$$

which yields the closed-loop system:

$$A_s\ddot{p} + \bar{B}_s\dot{p} + \bar{C}_s p = 0 \tag{37}$$

where $\bar{B}_s$ and $\bar{C}_s$ are as defined in (22), but $B_s, C_s, \Gamma, G_p, G_r$ are modified as stated previously. The following theorem gives sufficient conditions for the stability of this closed-loop syetem.

Theorem 2. *The closed-loop system of eq. (37) is stable in the sense of Lyapunov if $G_p > 0$ and $G_r \geq 0$. The system is asymptotically stable if $G_p > 0$ and $G_r > 0$.* ■

As in Theorem 1, this stability property holds regardless of the number of modes in the model, or the knowledge of the parameters.

We next consider controllers using velocity feedback.

Stabilization Using Total Velocity Feedback Controllers (TVFC)

During deployment, assembly or initial startup phase of many LFSS, the stability of the structure, and not the pointing performance, is of primary importance; that is, the elastic motion must be damped out, and the rigid-body attitude *rate* must be zero. The system must remain in a state of rest, although its orientation (attitude) in this state is not important. For such application, it is sufficient to use feedback of the total measured velocity, and to omit attitude feedback (i.e., $G_p = 0$). The resulting closed-loop system is given by eq. (37), where

$$\bar{B}_s = B_s + \Gamma^T G_r \Gamma \tag{38}$$

and

$$\bar{C}_s = C_s = \text{diag}[0, \Lambda_{n_q \times n_q}] \tag{39}$$

From (37), (38) and (39), it is clear that there is no "α" term in the closed loop equation (37). That is, the closed-loop system has six zero eigenvalues corresponding to the three rigid-body rotational angles (α). Recalling that the attitude is not important in this configuration, the "α" in eq. (37) can be removed, and the resulting system can be expressed in the following state space form:

$$\frac{d}{dt}\begin{bmatrix} q \\ \dot{\alpha} \\ \dot{q} \end{bmatrix} \begin{matrix} = \\ = \\ = \end{matrix} \begin{bmatrix} 0_{n_q} & 0_{n_q\times 3} & I_{n_q} \\ E_{11} & E_{12} & E_{13} \\ E_{21} & E_{22} & E_{23} \end{bmatrix} \begin{bmatrix} q \\ \dot{\alpha} \\ \dot{q} \end{bmatrix} \tag{40}$$

where E_{ij} are appropriately dimensioned constant matrices. The following theorem investigates the stability of eq. (40).

Theorem 3. *The closed-loop system with TVFC, given by eq. (40), is stable in the sense of Lyapunov if $G_r \geq 0$, and is asymptotically stable if $G_r > 0$.*

Proof. Consider the Lyapunov function

$$V(q, \dot{\alpha}, \dot{q}) = q^T \Lambda q + \dot{p}^T A_s \dot{p} \tag{41}$$

V is clearly positive definite. Taking the derivative,

$$\dot{V} = 2\dot{q}^T \Lambda \dot{q} + 2\dot{p}^T A_s \ddot{p} \tag{42}$$

Using (37) in (42) [with $\bar{B}_s$ and $\bar{C}_s$ as defined in (38) and (39)], we obtain:

$$V = -2\dot{p}^T \bar{B}_s \dot{p} = -2\dot{q}^T D\dot{q} - 2\dot{p}^T \Gamma^T G_r \Gamma \dot{p} \tag{43}$$

If $G_r \geq 0$, then $\dot{V} \leq 0$, and the system is Lyapunov-stable. Suppose $G_r > 0$. Then $\dot{V} = 0 \Rightarrow \dot{q} = 0$, and

$$\dot{\alpha} \mathcal{I}^T G_r \mathcal{I} \dot{\alpha} = 0 \tag{44}$$

where

$$\mathcal{I} = [I_3, \quad I_3, \quad \ldots, I_3]^T_{3\times 3m_T} \tag{45}$$

Since $\mathcal{I}$ is of full rank, $\dot{V} = 0 \Rightarrow \dot{\alpha} = 0$. From (37), (38) and (39), $\dot{\alpha} = 0$, $\dot{q} = 0$ implies $\Lambda q = 0$, i.e., $q = 0$. Thus $\dot{V}$ is not zero along any trajectories, and the system is asymptotically stable. ■

The TVFC accomplishes damping enhancement as well as attitude stabilization. That is (from Theorem 3), as $t \to \infty$, the elastic motion $q \to 0$ (generally at a rate faster than the open-loop case), and the rigid-body angular velocity $\dot{\alpha} \rightarrow 0$. Thus the attitude (i.e., the orientation) tends to a constant. This type of stability is sufficient during deployment, assembly, or construction of LFSS, and during the startup. For normal operation, however, feedback of the attitude must be included, as discussed previously (Theorem 2).

Damping Enhancement Using CDEC

The CDEC differs from the TVFC in that it leaves the rigid modes completely unaffected. Normally it is used as a "secondary" [Jos.80], or a "low-authority" [Aub.79] controller, and forms the inner loop of the complete control system, wherein the "primary", or "high-authority" controller is used in the outer loop to accomplish attitude control. Thus the function of the CDEC is only damping enhancement. It can be proved that the CDEC has the same robust stability property (with $G_r > 0$) as the CAC and the TVFC. The proof using the Lyapunov function: $V(q, \dot{q}) = q^T \Lambda q + \dot{q}^T \dot{q}$, is very similar, and is omitted.

2.3 Design of Dissipative Controllers

So far we have considered the stability properties of dissipative controllers. In this subsection we concentrate on how to obtain the gains, G_p and G_r. We first prove that the CAC minimizes a particular quadratic performance index. To be specific, every CAC is an optimal linear quadratic regulator (LQR), and minimizes a quadratic performance function which includes a state-control cross penalty in addition to the standard state and control penalties. We also prove that every CDEC is an optimal LQR which minimizes a quadratic performance function with no cross penalty. These facts provide some insight into the design of dissipative controllers. We next formulate the CDEC design problem as an optimal output feedback problem, and obtain the necessary conditions for minimizing a stochastic quadratic performance function. Finally, we also present an alternate design method which uses the least squares solution to obtain the desired pole locations.

Theorem 4. *For the system given by (4), the CAC given by (14)-(16) minimizes the performance function*

$$J = \int_0^\infty (x^T Q x + 2x^T S u + u^T R u) dt \tag{46}$$

where

$$x = (p^T, \dot{p}^T)^T \tag{47}$$

and

$$Q = \begin{bmatrix} \Gamma^T G_p G_r^{-1} G_p \Gamma & 0 \\ 0 & \Gamma^T G_r \Gamma + 2B_s \end{bmatrix}, \quad R = G_r^{-1}, S = \begin{bmatrix} \Gamma^T G_p G_r^{-1} \\ 0 \end{bmatrix} \tag{48}$$

Proof. Defining x as in (47), (4) can be expressed as:

$$\dot{x} = Ax + Bu \tag{49}$$

where

$$A = \begin{bmatrix} 0 & I \\ -A_s^{-1} C_s & -A_s^{-1} B_s \end{bmatrix}, \quad B = \begin{bmatrix} 0 \\ A_s^{-1} \Gamma^T \end{bmatrix} \tag{50}$$

From [Ath.66] the control law which minimizes J in (46) is: given by: $u = Gx$, where

$$G = -R^{-1}[B^T P + S^T] \tag{51}$$

$$A^T P + PA - SR^{-1} B^T P - PBR^{-1} S^T - PBR^{-1} B^T P = -(Q - SR^{-1} S^T) \tag{52}$$

Upon substitution for A and B from (50) into (52), it can be verified that

$$P = \begin{bmatrix} C_s + \Gamma^T G_p \Gamma & 0 \\ 0 & A_s \end{bmatrix} \tag{53}$$

Using (53) in (51), we have:

$$G = -[G_p, \quad G_r] \quad \begin{bmatrix} \Gamma & 0 \\ 0 & \Gamma \end{bmatrix} \tag{54}$$

Thus the control law (16) solves an optimal LQR problem. ∎

The performance index in (46) can be expressed as:

$$J = \int_0^\infty \left[y_p^T G_p G_r^{-1} G_p y_p + y_r^T G_r y_r + 2\dot{q}^T D \dot{q} + 2 y_p^T G_p G_r^{-1} u + u^T G_r^{-1} u \right] dt$$

This expression shows the penalties on attitude, attitude rate, elastic motion, and the control vector.

The following theorem states a similar result for the CDEC.

Theorem 5. *For the system given by (19), the CDEC control law:* $u = -G_r \tilde{y}_r$, *where* $\tilde{y}_r$ *is as in (20), minimizes the performance function:*

$$J = \int_0^\infty \left[x^T Q x + u^T R u \right] dt \tag{55}$$

where

$$x = \left(q^T, \dot{q}^T \right)^T \tag{56}$$

$$Q = \begin{bmatrix} 0 & 0 \\ 0 & 2D + \tilde{\Gamma}^T G_r \tilde{\Gamma} \end{bmatrix} \quad \text{and} \quad R = G_r^{-1} \tag{57}$$

∎

The proof is straightforward and is similar to that of Theorem 4. It can be verified that the Riccati matrix is given by:

$$P = \begin{bmatrix} \Lambda & 0 \\ 0 & I \end{bmatrix} \tag{58}$$

The result in Theorem 5 for the CDEC was originally proved in [Arb.81].

It can be shown in an entirely analogous manner that the TVFC also minimizes a quadratic performance function with no state- control cross penalty.

The results of Theorems 4 and 5 give the performance functions which are minimized by the dissipative controllers, and provide a clue for selecting the gains G_p and G_r. For example, one method for selecting the gain G_r for the CDEC would be to first choose a state weighting matrix Q_d and then to select G_r which minimizes the distance between Q_d and Q (of Eq. (57)). However, the choice of Q_d is inherently limited because of the structure of Q in (57). In particular, Q_d must be of the form: $Q_d = \text{diag}(0, W)$; that is, it allows the weighting of $\dot{q}$ only. With this constraint, the G_r which minimizes the trace of $(Q_{22} - W)^2$ is given by [Arb.81]:

$$G_r = (\Gamma\Gamma^T)^{-1}\Gamma W \Gamma^T (\Gamma\Gamma^T)^{-1}$$

Similar results can be obtained for the CAC. However, the choice of the "desired" weighting matrices would be inherently limited because of the structure of Q and S.

Another approach to selecting G_r for the CDEC is to formulate the problem as an optimal linear quadratic output feedback problem. In order to retain the guaranteed stability property, it is necessary to constrain G_r to be *positive definite* and *symmetric*, which would cause difficulty in deriving the necessary conditions for optimality. This problem can be avoided if G_r is constrained to be *diagonal*, which is true if the CDEC is in the "member damper" mode [Can.78].

Following [Jos.80b], the problem is then posed as follows:

Given the system

$$\dot{x} = Ax + Bu + v \tag{59}$$

$$z = Cx + w \tag{60}$$

$$u = Gz \tag{61}$$

where v and w are white noise processes with covariance intensities V and W, and

$$A = \begin{bmatrix} 0 & I \\ -\Lambda & -D \end{bmatrix} \qquad B = \begin{bmatrix} 0 \\ \tilde{\Gamma}^T \end{bmatrix} \qquad C = B^T \tag{62}$$

find the positive diagonal gain matrix G, which will minimize the performance function:

$$J = \lim_{t_f \to \infty} \frac{1}{t_f} \mathcal{E} \int_0^{t_f} [x^T Q x + (GCx)^T R (GCx)] dt \tag{63}$$

("$\mathcal{E}$" denotes the expectation operator). Note that only the deterministic part of "u" is weighted in (63), because the variance of the noise-dependent part "GCw" is infinite.

This optimal output feedback problem can be solved using the procedure of [Jos.75]. Let the symbol $L * M$ denote the element by element product of two similarly dimensioned matrices, i.e.,

$$[L * M]_{ij} = L_{ij} M_{ij} \tag{64}$$

Define the $\ell \times 1$ vector function of an $\ell \times \ell$ matrix as:

$$\Delta(L) = \begin{bmatrix} L_{11} \\ L_{22} \\ \cdot \\ \cdot \\ \cdot \\ L_{\ell\ell} \end{bmatrix} \tag{65}$$

The following theorem gives the necessary conditions for optimality:

Theorem 6. *The necessary conditions for the minimization of J in (63) subject to the constraints of (59)-(61) are given by:*

$$g = -\left[R * (B^T \Sigma B) + W * (B^T P B)^{-1}\right] \Delta(B^T \Sigma P B) \tag{66}$$

$$(A + BGB^T)^T P + P(A + BGB^T) + Q + BGRGB^T = 0 \tag{67}$$

$$(A + BGB^T)\Sigma + \Sigma(A + BGB^T)^T + V + BGWGB^T = 0 \tag{68}$$

where P and Σ are $2n_q \times 2n_q$ symmetric matrices, and $g = (g_1, g_2, .., g_m)^T$ is the vector consisting of the diagonal elements of G.

Proof. The necessary conditions are obtained by expressing J as:

$$J = Tr[(Q + BGRGB^T)\Sigma]$$

where $\Sigma = \lim_{t\to\infty} \mathcal{E}[x(t)x^T(t)]$. The facts that $C = B^T$ and G is diagonal, are also utilized.

The proof is very similar to that for the general optimal output feedback problem [Jos.75]. The only difference is that the derivative of the Hamiltonian with respect to the vector g (rather than the matrix G is equated to zero). We also use the following properties of the trace of a matrix:

$$\frac{\partial}{\partial g} Tr[G\alpha G\beta] = \frac{\partial}{\partial g}[g^T(\alpha * \beta)g] = [\alpha * \beta + (\alpha * \beta)^T]g \tag{69}$$

$$\frac{\partial}{\partial g} Tr[G\alpha] = \Delta(\alpha) \tag{70}$$

■

The optimal $g*$ can be computed using the algorithm in [Jos.75], which is as follows:

1) choose an initial g with all positive elements.
2) solve the Lyapunov equation (67) for P
3) solve (66) and (68) simultaneously for G and Σ.
4) If G has converged, stop. If not, go to step 2.

This algorithm guarantees successive reduction in J, but the simultaneous solution of (66) and (68) presents a difficult numerical problem. An alternative approach would be to employ numerical algorithms such as Davidon-Fletcher-Powell (e.g., [Dav.68]). In our experience, such numerical minimization algorithms have been found to be very effective. The resulting controller is optimal (at least locally), and also has guranteed stability.

These results can be extended in a similar manner to the CAC. However, in general, optimal output feedback problems suffer from the existence of many local minima, as well as slow convergence. In addition, numerical conditioning plays an important role in the minimization algorithms.

Another method for selecting the feedback gains is based on pole location. In particular, we attempt to make the closed-loop matrices $\bar{B}_s$ and $\bar{C}_s$ equal certain "desired" values, B_d and C_d. For example, for the complete (rotational and translational) LFSS control problem, suppose we want to assign closed-loop damping ratios and frequencies ρ_α and ω_α to the rigid rotational modes and damping ratio ρ_{di} to the ith structural mode $(i = 1, 2, .., n_q)$ without changing its natural frequency (ω_i). Then we define

$$B_d = 2\,\mathrm{diag}\left(\rho_\alpha \omega_\alpha J_s, \rho_{d1}\omega_{d1}, \rho_{d2}\omega_{d2}, ..., \rho_{dnq}\omega_{dn_q}\right) \tag{71}$$

$$C_d = \mathrm{diag}\left(\omega_\alpha^2 J_s, \omega_1^2, \omega_2^2, ..., \omega_{n_q}^2\right) \tag{72}$$

From (22), (71) and (72), we have

$$\Gamma^T G_r \Gamma = B_d - B_s \triangleq \bar{B}_d \tag{73}$$

$$\Gamma^T G_p \Gamma = C_d - C_s \triangleq \bar{C}_d \tag{74}$$

We assume that Γ is of full rank, or has been reduced to be of full rank by eliminating sensor/actuator locations corresponding to linearly dependent rows. If it is desired to control only n_c, modes $(n_c < n_q + 3)$, then the rows corresponding to the uncontrolled modes are assumed to be removed from Γ^T. We also assume that $\bar{B}_d$ and $\bar{C}_d$ are positive-definite. If the number of controlled modes (including the three rigid modes), n_c, is less than the number of actuators, m, many solutions for (73) exist, and the solution

$$G_r = \Gamma(\Gamma^T\Gamma)^{-1}\bar{B}_d(\Gamma^T\Gamma)^{-1}\Gamma^T \tag{75}$$

minimizes the Frobenius norm of G_r, which is defined as:

$$\|G_r\|_F = \left(\sum_i \sum_j G_{r_{ij}}^2\right)^{1/2} \tag{76}$$

For the case where $n_c > m$, no solution exists, and the least squares solution (which minimizes $\|B_d - \bar{B}_s\|_F$) is given by [Ell.79]:

$$G_r = (\Gamma\Gamma^T)^{-1}\Gamma\bar{B}_d\Gamma^T(\Gamma\Gamma^T)^{-1} \tag{77}$$

(For $n_c = m$, of course, a unique solution exists).

The expressions for G_p are entirely similar. Since these solutions always give positive definite G_p and G_r, the resulting CAC is guaranteed to be stable. Just as in the case of the optimal output feedback method, this method relies on the knowledge of the system parameters, that is, frequencies, damping ratios and mode shapes and slopes. However, because of the constraints imposed on the control law under this (collocated) framework, the stability is always guranteed regardless of the inaccuracies in the parameter values used in the design; only the optimality may be degraded because of the lack of accurate parameter knowledge.

2.4 Robustness Properties of Dissipative Controllers

We have established in Sec. 2.2 that, with perfect (i.e., linear and instantaneous) sensors and actuators, the stability of dissipative controllers is guranteed regardless of the number of modes controlled (or even modeled), and regardless of parameter errors. This stability robustness property makes this class of controllers very attractive. This property, although highly desirable, is not adequate, because the stipulated ideal conditions (i.e., exact actuator/sensor collocation, perfect actuators and sensors) do not exist in practice. In this section, we consider some practically motivated problems encountered in the implementation of dissipative controllers, and investigate if the robust stability is still maintained. In particular, we investigate the robustness of such controllers to: 1) imprecise collocation of sensors and actuators, 2) actuator/sensor imperfections including nonlinearities, and finite bandwidth.

2.4.1 Robustness to Imprecise Collocation

From a practical engineering viewpoint, it may not be possible to place a sensor and an actuator precisely at the same location. Therefore, the mode shapes and slopes at the sensor and actuator locations will not be equal; that is, the "r_k", "Ψ_k" and

"Φ_k" in the sensor equations (11) and (12) will be different from those in the control influence matrix in the system equation (8). Suppose the translational and rotational displacements are given by:

$$y_p = (\Gamma + \delta\Gamma_1)p$$

$$y_r = (\Gamma + \delta\Gamma_2)\dot{p}$$

where

$$\delta\Gamma_1 = \begin{bmatrix} 0 & \delta\tilde{R}_1 & \delta\Psi_1 \\ 0 & 0 & \delta\Phi_1 \end{bmatrix} \triangleq [0\,\Delta_1] \tag{78}$$

$$\delta\Gamma_2 = \begin{bmatrix} 0 & \delta\tilde{R}_2 & \delta\Psi_2 \\ 0 & 0 & \delta\Phi_2 \end{bmatrix} \triangleq [0\,\Delta_2] \tag{79}$$

$\delta\Psi_1, \delta\Phi_1, \delta\Psi_2, \delta\Phi_2$ are the differences in the mode shape and slope matrices at the position and rate sensors, and $\delta\tilde{R}_1, \delta\tilde{R}_2$ are the errors in cross product matrices because of imprecise collocation of force actuators and position and rate sensors. As proved in Theorem 1, with positive definite position and rate gains, the closed-loop system is asymptotically stable (a.s.) with perfect collocation. Denoting the closed loop matrix under ideal conditions (perfect collocation) by A_c, given a positive definite $n \times n$ matrix Q, there exists a positive definite $n \times n$ matrix P such that

$$A_c^T P + P A_c = -Q \tag{80}$$

where

$$A_c = A - BG \tag{81}$$

where A, B, and G are given by Eqs. (50) and (54). The following theorem gives bounds on Δ_1 and Δ_2 which ensure stability.

Theorem 7. *The closed-loop system with imprecisely collocated sensors and actuators is a.s. if*

$$\|G_p\|_s\|\Delta_1\|_s + \|G_r\|_s\|\Delta_2\|_s < \frac{\lambda_m(Q)\lambda_m(A_s)}{2\|\Gamma\|_s\lambda_M(P)} \tag{82}$$

where $\|L\|_s$ denotes the spectral norm of a matrix L:

$$\|L\|_s = \left[\lambda_M(L^T L)\right]^{1/2} \tag{83}$$

and $\lambda_M()$ and $\lambda_m()$ denote the largest and the smallest eigenvalues.

Proof. The closed-loop system is given by:

$$\dot{x} = A_c x - \begin{bmatrix} 0 & 0 \\ A_s^{-1}\Gamma^T G_p \delta\Gamma_1 & A_s^{-1}\Gamma^T G_r \delta\Gamma_2 \end{bmatrix} x \triangleq (A_c - E)x \tag{84}$$

Consider the Lyapunov function:

$$V(x) = x^T P x \tag{85}$$

Then it can be shown that

$$\dot{V} = -x^T Q x + 2x^T P E x \tag{86}$$

For $\dot{V}$ to be negative definite, we need $x^T PEx < x^T Qx$. Since

$$x^T PEx \leq \|P\|_s\|E\|_s\|x\|^2 \quad \text{and}$$

$$x^T Q x \geq \lambda_m(Q)\|x\|^2$$

and noting that $\|P\|_s = \lambda_M(P)$, $\dot{V}$ is negative definite if

$$\|E\|_s \leq \lambda_m(Q)/2\lambda_M(P) \tag{87}$$

The inequality in (82) can be obtained by using the properties of the spectral norm to obtain an upper bound on $\|E\|_s$, i.e., by noting that

$$\|E\|_s \leq \|A_s^{-1}\Gamma^T G_p \delta\Gamma_1\|_s + \|A_s^{-1}\Gamma^T G_r \delta\Gamma_2\|_s \leq \lambda_m^{-1}(A_s)[\|G_p\|_s\|\Delta_1\|_s + \|G_r\|_s\|\Delta_2\|_s]\|\Gamma\|_s$$

■

It should be noted that the bound obtained in Theorem 7 is generally conservative, and depends on the knowledge of the parameters. It also depends on the choice of Q. However, it is apparent from the theorem that the system will be a.s. for sufficiently small collocation errors.

2.4.2 Robustness to Actuator/Sensor Nonlinearities and Dynamics

Although dissipative controllers have attractive stability properties with perfect (i.e., linear, instantaneous) sensors and acuators, the devices available in practice have nonlinearities and phase lags. In order to be useful in practical applications, the controller should be tolerant to nonlinearities (e.g., saturation, relays, deadzones, etc.), and to phase shifts (e.g. actuator dynamics and/or computational delays). Uncertainties usually exist in the knowledge of the nonlinearities and the phase lags. For these reasons, we shall investigate the closed-loop stability of dissipative controllers in the presence of unmodeled sensor/actuator dynamics and nonlinearities. In particular, we will prove that the robust stability property of dissipative controllers is still valid in the presence of a wide class of actuator and sensor imperfections. In particular, we will prove that the CAC preserves global asymptotic stability when linear actuator dynamics satisfying certain phase conditions are present, or monotonic increasing actuator nonlinearities and sensor nonlinearities belonging to the $(0,\infty)$ sector are present. [A function $\psi(\nu)$ is said to belong to the $(0,\infty)$ sector if $\psi(0) = 0$ and $\nu\psi(\nu) > 0$ for $\nu \neq 0$; ψ is said to belong to the $[0,\infty)$ sector if $\nu\psi(\nu) \geq 0$ (Fig. 1)]. The stability with velocity feedback controllers (CDEC and TVFC) is proved under much weaker conditions. In particular, such controllers have at least 90 degrees phase margin and can tolerate actuator and sensor nonlinearities belonging to the $(0,\infty)$ sector ($[0,\infty)$ sector for CDEC). The re-

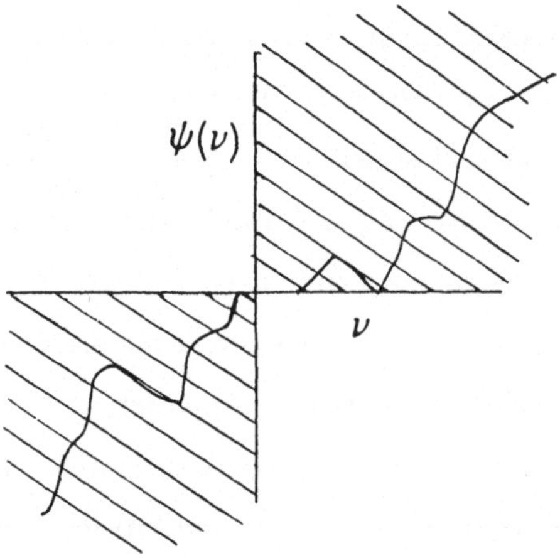

Figure 1. Nonlinearity belonging to [0, ∞) sector

sults significantly enhance the viability of both types of controllers, especially when the avilable information about the LFSS parameters is inadequate or inaccurate.

For the case where the actuators are represented by an operator $\mathcal{H}$, The actual input $u(t)$ is given by:

$$u(t) = \mathcal{H} u_c(t) \tag{88}$$

where u_c is the ideal (desired) input, $\mathcal{H}$ is a nonanticipative, linear or nonlinear, time-varying or invariant operator. For CAC, it is proved that the closed-loop system is globally asymptotically stable if either

1) $\mathcal{H}$ is linear, time-invariant (LTI) and stable with a rational transfer matrix $H(s)$ which satisfies certain frequency-domain conditions.
2) The actuators have time-invariant, monotonic increasing nonlinearities and sensors have nonlinearities belonging to the $(0, \infty)$ sector.

For CDEC, it is proved that global asymptotic stability is preserved when

1) $\mathcal{H}$ is a stable nonlinear dynamic operator and satisfies certain passivity conditions, or
2) $\mathcal{H}$ is a stable LTI operator with phase within $\pm 90^o$
3) The sensors and actuators have time-varying nonlinearities belonging to the $[0, \infty)$ sector.
4) $\mathcal{H}$ consists of stable first-order dynamics followed by time-invariant nonlinearities belonging to the $[0, \infty)$ sector.

Similar results are presented for the TVFC, but with the $[0, \infty)$ sector replaced by the $(0, \infty)$ sector. These analytical results significantly enhance the stability and robustness properties of dissipative controllers, and therefore increase their practical applicability. The methods used in the proofs include Lyapunov and function-space methods, and are similar to those given in [Des.75] and [Pop.73].

The linearized mathematical model of an LFSS was given by eqs. (4)- (15). As stated previously, the center of mass translation is not important in the attitude control problem. For this reason, and to keep the derivations simpler, we consider only the rotational rigid-body motion, and the elastic motion, of the LFSS. We utilize torque actuators and attitude and rate sensors (e.g., start trackers and rate gyros). The mathematical model is then given by eqs. (27)-(33). Thus the sensor outputs (ignoring noise) are given by eqs. (34) and (35). Although we consider only the rotational motion, all the results proved in this subsection would still be valid for the original system which includes the rigid c.m. translation.

As stated previously, D is assumed to be a symmetric positve definite matrix which represents the inherent structural damping. The control law for the CAC is given by:

$$u_c = u_{cp} + u_{cr} \tag{89}$$

where

$$u_{cp} = -G_p y_p \tag{90}$$

$$u_{cr} = -G_r y_r \tag{91}$$

where u_c represents the command input, u_{cp} and u_{cr} represent command attitude and rate input, and G_p, G_r are $m \times m$ feedback gain matrices (where $m = 3m_T$).

For CDEC, the rigid-body rates are removed from the feedback signal by subtracting attitude rates at two locations. Consequently, the model used for damping enhancement is given by (19) and (20). The control law for CDEC is given by:

$$u_c = -G_r \tilde{y}_r \tag{92}$$

For TVFC, the control law is given by (89)-(91), but with $G_p = 0$.

The problem considered in this section is to investigate the closed-loop stability for the case where the actuators and sensors are represented by dynamic operators and/or nonlinear gains. Our approach is to make use of input-output stability concepts and the Lyapunov method. We start by defining the terminology and the concepts, which are adopted from [Des.75].

Mathematical Preliminaries

Consider the linear vector space L_2^n of real square-integrable n-vector functions of time t, defined as:

$$L_2^n = \{g : R_+ \to R^n \mid \int_0^\infty g^T(t)g(t)dt < \infty\} \tag{93}$$

where R^n is the linear space of ordered n-tuples of real numbers, and R_+ denotes the interval $0 \leq t < \infty$. The scalar product is defined as:

$$< g_1, g_2 >= \int_0^\infty g_1^T(t)g_2(t)dt \tag{94}$$

For $g \varepsilon L_2^n$, its norm is defined as

$$\|g\| =< g, g >^{1/2} \tag{95}$$

Define the truncation operator P_T such that

$$\begin{aligned} g_T(t) = P_T g(t) &= g(t) \quad 0 \leq t \leq T \\ &= 0 \quad t > T \end{aligned} \tag{96}$$

Define the extended space L_{2e}^n:

$$L_{2e}^n = \{g : R_+ \to R^n | g_T \varepsilon L_2^n \;\; \forall \;\; T \geq 0\} \tag{97}$$

Thus L_{2e}^n is a linear vector space of functions of t whose truncations are square-integrable on $[0, T)$ for all $T < \infty$. For $g_1, g_2 \,\varepsilon\, L_{2e}^n$, define the truncated inner product

$$< g_1, g_2 >_T =< g_{1T}, g_{2T} >= \int_0^T g_1^T(t)g_2(t)dt \tag{98}$$

The truncated norm is defined by: $\|g\|_T =< g, g >_T^{1/2}$.

Consider an operator $\mathcal{H} : L_{2e}^n \rightarrow L_{2e}^n$. $\mathcal{H}$ is said to be strictly passive iff there exist finite constants β and $\delta > 0$ such that

$$< \mathcal{H}g, g >_T \geq \beta + \delta \|g\|_T^2 \quad \forall \; T > 0, \forall g \varepsilon L_{2e}^n \tag{99}$$

$\mathcal{H}$ is passive if $\delta \geq 0$ in (99).

Robustness of Collocated Attitude Controllers (CAC)

Stability With Dynamic Operator in the Loop- Suppose $\mathcal{H}$ is a finite- dimensional, linear, time-invariant (LTI)operator. [The sensors usually have high bandwidth, and $\mathcal{H}$ then represents the actuators (Fig. 2). For the case where $\mathcal{H}$, G_p and G_r are diagonal, $\mathcal{H}$ represents the combined sensor/actuator dynamics.] We denote $\mathcal{H}g$ by $\mathcal{H}(z_0; g)$ where z_0 is the initial state vector of $\mathcal{H}$.

Theorem 8. *Suppose $\mathcal{H}$ is a non-anticipative, asymptotically stable, observable, LTI operator whose transfer matrix is $H(s) = \varepsilon I + \hat{H}(s)$, where $\varepsilon > 0$ and $\hat{H}(s)$ is a proper, minimum-phase, rational matrix. Under these conditions, the closed-loop system given by Eqs. (27), (34), (35), and (88)-(91) is asymptotically stable (a.s.) if*

$$\hat{H}(j\omega)\left(\omega^2 G_r - jG_p\omega\right) + \left(\omega^2 G_r + jG_p\omega\right)\hat{H}^*(j\omega) \geq 0 \quad \text{for all real } \omega. \tag{100}$$

where $*$ denotes the conjugate transpose.

Proof. Define the function

$$V(t) = p^T C_s p + \dot{p}^T A_s \dot{p} \tag{101}$$

$C_s \geq 0, A_s > 0$, implies $V(t) \geq 0$ for all $t \geq 0$. Differentiating V with respect to t, and using (27), (34), (35), (88)-(91),

$$\dot{V} = -2\dot{p}^T B_s \dot{p} - 2u_{cr}^T G_r^{-1} \mathcal{H}[z_0; u_c] \tag{102}$$

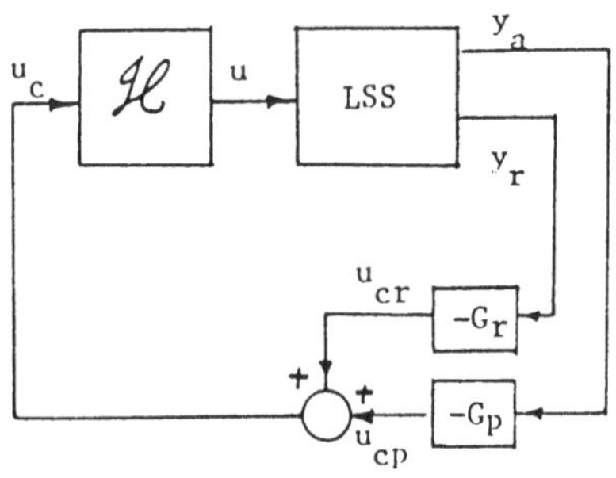

Figure 2. Dissipative controller with dynamic operator in the loop

where $\mathcal{H}$ also depends on its initial state z_0. Since $\mathcal{H}$ is linear,

$$\mathcal{H}[z_0; u_c] = h_0(t) + \mathcal{H}[0; u_c] \tag{103}$$

where $h_0(t)$ is the unforced response of $\mathcal{H}$ due to nonzero initial state. Since $\mathcal{H}$ is strictly stable, $\|h_0\|$ is finite for any finite z_0.

Substituting (103) in (102) and integrating from 0 to T, since $V(T) \geq 0$,

$$\begin{aligned} 0 \leq V(T) = V(0) - 2 < \dot{p}, B_s\dot{p} >_T - 2 < u_{cr}, G_r^{-1}h_0 >_T \\ - 2 < u_{cr}, G_r^{-1}\mathcal{H}_p u_{cp} >_T \end{aligned} \tag{104}$$

where u_{cp} and u_{cr} are as defined in (90) and (91), and

$$\mathcal{H}_p u_{cp} = \mathcal{H}[0; (I + sG_rG_p^{-1})u_{cp}] \tag{105}$$

In (105), "s" denotes the derivative operator. (s is technically noncausal; however, this difficulty can be overcome by defining the derivative of a truncation at T to be equal to that of the untruncated function.) Using Parseval's theorem,

$$\begin{aligned} < u_{cr}, G_r^{-1}\mathcal{H}_p u_{cp} >_T &= \frac{1}{2\pi}\int_{-\infty}^{\infty} U_{cr_T}^*(j\omega)G_r^{-1}H(j\omega)(I + j\omega G_rG_p^{-1})U_{cp_T}(j\omega)d\omega \\ &= \frac{1}{2\pi}\int_{-\infty}^{\infty} U_{cr_T}^*(j\omega)G_r^{-1}H(j\omega)\left(\frac{G_P}{j\omega} + G_r\right) G_r^{-1}U_{crT}(j\omega)d\omega \\ &= \frac{1}{2\pi}\int_{-\infty}^{\infty} U_{cr_T}^*(j\omega)G_r^{-1}\left[H(j\omega)\left(\frac{G_p}{j\omega} + G_r\right)\right. \\ &\quad + \left.\left(\frac{G_p}{-j\omega} + G_r\right)H^*(j\omega)\right] G_r^{-1}U_{cr_T}(j\omega)d\omega \end{aligned} \tag{106}$$

The matrix in the brackets is positive (from Eq. 100), and we have

$$< u_{cr}, G_r^{-1}\mathcal{H}_p u_{cp} >_T \geq \varepsilon_1\|u_{cr}\|_T^2 \tag{107}$$

where $\varepsilon_1 = \varepsilon/\lambda_M(G_r)$. This yields [from (104)]

$$0 \leq V(0) - 2 < \dot{q}, D\dot{q} >_T - 2\varepsilon_1 \|u_{cr}\|_T^2 - 2 < u_{cr}, G_r^{-1} h_0 >_T \tag{108}$$

wherein we have used the fact that $\dot{p}^T B_s \dot{p} = \dot{q}^T D \dot{q}$. Therefore,

$$\lambda_m(D)\|\dot{q}\|_T^2 + \varepsilon_1 \|u_{cr}\|_T^2 \leq V(0)/2 + \|u_{cr}\|_T \|G_r^{-1}\|_s \|h_0\| \tag{109}$$

where $\|\|_s$ denotes the spectral norm of a matrix. Eq. (109) can be written as

$$\lambda_m(D)\|\dot{q}\|_T^2 + [c_1\|u_{cr}\|_T - (c_2/2c_1)]^2 \leq V(0)/2 + c_2^2/4c_1 \tag{110}$$

where $c_1 = \sqrt{\varepsilon_1}$ and $c_2 = \|h_0\|$. Therefore,

$$\lim_{t\to\infty} \dot{q}(t) = 0, \text{and} \lim_{t\to\infty} u_{cr}(t) = 0.$$

Denoting the rigid-body attitude $\alpha = (\phi, \theta, \psi)^T$, this implies that

$$\lim_{t\to\infty} \dot{\alpha}(t) = 0.$$

Taking the limit of the closed-loop equation as $t \to \infty$,

$$\begin{bmatrix} 0 \\ \\ \Lambda\bar{q} \end{bmatrix} = \begin{bmatrix} I \\ \\ \Phi^T \end{bmatrix} \overline{\mathcal{H} u_{cp}} \tag{111}$$

where the overhead bar denotes the limit as $t \to \infty$. From (111), $\overline{\mathcal{H} u_{cp}} = 0$ and $\bar{q} = 0$, which yields $\bar{\alpha} = 0$. Since $\mathcal{H}$ is observable and minimum phase and its output tends to zero, its state vector tends to zero as $t \to \infty$, and the system is asymptotically stable. ■

The following corollary essentially states that, for diagonal G_p, G_r and H, it is sufficient that the phase lag of $\hat{H}(j\omega)$ is less than the phase lead introduced by the controller.

Corollary 8.1. *Suppose G_p, G_r, and H are diagonal and satisfy conditions of Theorem 8. Then the closed-loop system is asymptotically stable if, for $i = 1, 2, ..., m$,*

$$-\arctan(\omega G_{ri}/G_{pi}) < \arg[\hat{H}_i(j\omega)] \le 180^o - \arctan(\omega G_{ri}/G_{pi}) \quad \text{for all real} \quad \omega \tag{112}$$

where arg( ) denotes the phase angle of a complex variable, and the subscript i denotes the ith diagonal element. ■

For the case where $\hat{H}_i(s) = k_i/(s + a_i)$, with $k_i, a_i > 0$, condition (112) becomes

$$G_{ri}/G_{pi} \ge 1/a_i \tag{113}$$

Thus, for the case where the sensors are perfect (high bandwidth) and the actuators have first-order dynamics, the system is asymptotically stable if the ratio of rate to proportional gain is at least equal to the inverse of the actuator pole.

In Theorem 8 and Corollary 8.1, the transfer function of $\mathcal{H}$ was assumed to be of the form: $H(s) = \varepsilon I + \hat{H}(s)$, where $\varepsilon > 0$. That is, a direct transmission term, no matter how small, was present. From Theorem 1, the closed-loop system is a.s. for any $\varepsilon > 0$. Therefore, the closed-loop eigenvalues are all in the open left half-plane. Because of continuity, it is obvious that, when $\varepsilon = 0$, the eigenvalues will not cross the imaginary axis. That is, the eigenvalues will be in the closed left half-plane. Theorem 9 given below considers the case when $\varepsilon = 0$. It essentially shows that, if the closed-loop system with no elastic modes in the loop is a.s., then so is the system with elastic modes, provided that (100) is satisfied with H replacing $\hat{H}$.

Theorem 9. *Suppose $\mathcal{H}$ is a non-anticipative, asympotically stable, observable, LTI operator with rational transfer matrix H(s) which is proper and minimum-phase. If the closed-loop system for the rigid body model alone (i.e., Eqs. (27), (34), (35) (88)-(91) with $n_q = 0$) is a.s., then the entire closed-loop system (i.e.,with $n_q \neq 0$) is a.s. if*

$$H(j\omega)\left(\omega^2 G_r - jG_p\omega\right) + \left(\omega^2 G_r - jG_p\omega\right) H^*(j\omega) \ge 0 \ \text{ for all real } \ \omega. \tag{114}$$

Proof. Considering the rigid-body equations,

$$J_s\ddot{\alpha} = \mathcal{H}u_c = \mathcal{H}(u_\alpha + u_q) \tag{115}$$

where $u_\alpha = -G_p\alpha - G_r\dot{\alpha}$ and $u_q = -G_p\Phi q - G_r\Phi\dot{q}$ Thus the transfer function from $\dot{q}$ to $\dot{\alpha}$ is given by

$$M(s) = [I + H(s)\{G_p + sG_r\}]^{-1}H(s)\{G_p + sG_r\}\Phi \tag{116}$$

Since the closed-loop rigid-body system is strictly stable by assumption, $M(s)$ is strictly stable and finite-gain, which implies

$$\|\dot{\alpha}\|_T \leq r\|\dot{q}\|_T + \|h_m\|_T \tag{117}$$

where r is the gain of M and h_m is its free response. Proceeding as in the proof of Theorem 8, we can arrive at Eq. (109) wherein $\varepsilon_1 = 0$ and h_0 is replaced by h_m. Since $u_{cr} = -G_r(\dot{\alpha} + \Phi\dot{q})$, we have from (117),

$$\|u_{cr}\|_T \leq c_1\|\dot{q}\|_T + c_2\|h_m\|_T \tag{118}$$

where c_1 and c_2 are positive constants. Completing squares as in (110) and noting that $\|h_m\|$ is finite, it can be proved that $\|\dot{q}\|_T$ is bounded for all $T > 0$, and that $\lim_{t\to\infty}\dot{q}(t) = 0$. From (118), u_{cr} also tends to zero as $t \to \infty$. The remainder of the proof is similar to that of Theorem 9. ■

Corollary 9.1. *Under the conditions of Theorem 9, if G_p, G_r, H are diagonal, then the closed-loop system is a.s. if (112) is satisfied with H replacing $\hat{H}$.* ■

From Corollary 9.1, for the case where $H_i(s) = k_i/(s + a_i)$ with $k_i, a_i > 0$, the closed-loop asymptotic stability is assured if $G_{pi} \leq a_iG_{ri}$ for $i = 1, 2, ..., m$.

The significance of the above result is that the stability can be assured by making the ratio of the rate to proportional gains sufficiently large. One has to know only the sensor/actuator characteristics; *the knowledge of the plant parameters is not required.* This result is completely consistent with the result obtained in [Jos.81] for single-input, single-output systems, for small G_p and G_r, using a root-locus argument.

We next consider the case where nonlinearities are present in the loop.

Stability in the Presence of Nonlinearities- Suppose the sensors and actuators have nonlinearities as shown in Figure 3. In this case, the actual input is given by:

$$u = \psi_a[-G_p\psi_p(y_p) - G_r\psi_r(y_r)] \tag{119}$$

where ψ_a, ψ_p, and ψ_r denote the actuator nonlinearity and the attitude and rate sensor nonlinearities, respectively. Assuming for simplicity that $m_T = 1$ (one three-axis actuator/sensor), and that G_p, G_r are diagonal,

$$u_i = \psi_{ai}[-G_{pi}\psi_{pi}(-w_i) - G_{ri}\psi_{ri}(-\dot{w}_i)] \tag{120}$$

where $w = -y_p$ (3×1 vector). We asume that ψ_{ai}, ψ_{pi}, and ψ_{ri} (i=1,2,3) are continuous single-valued functions: $R \to R$. The following theorem gives sufficient conditions for stability.

Theorem 10. *Consider the closed-loop system given by Eqs.(27), (34), (35), (88)-(91), and (119), where G_p and G_r are diagonal with positive entries. Suppose ψ_{ai}, ψ_{pi}, andψ_{ri} are single-valued continuous functions, and that, for i=1, 2, 3,*

(i) $\psi_{ai}(0)$ =0, ψ_{ai} are time-invariant and monotonic increasing (respectively, nondecreasing and nonzero in a neighborhood of the origin).

(ii) ψ_{pi}, ψ_{ri}, belong to the $(0, \infty)$ sector (respectively, to the $(0, \infty)$ and $[0, \infty)$ sectors) and ψ_{pi} are time-invariant.

Then the closed-loop system is globally asymptotically stable (respectively, stable in the sense of Lyapunov).

Proof. Define

$$\bar{\psi}_{pi}(\nu) = -\psi_{pi}(-\nu) \tag{121}$$

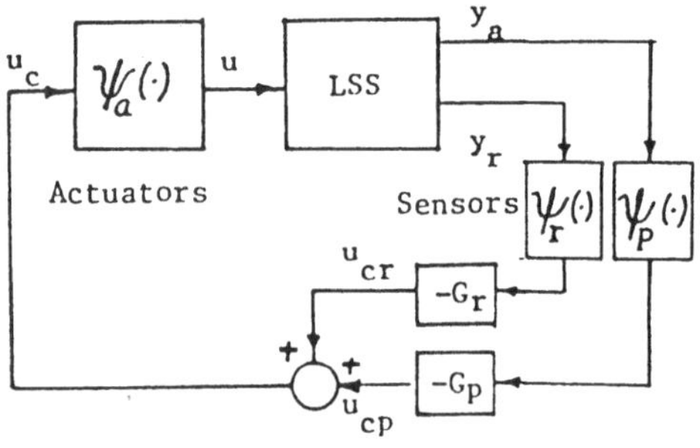

Figure 3. Nonlinearities in sensors and actuators

$$\bar{\psi}_{ri}(\nu) = -\psi_{ri}(-\nu) \tag{122}$$

If $\psi_{pi}, \psi_{ri} \,\epsilon\, (0,\infty)$ or $[0,\infty)$ sector, it is straightforward to show that $\bar{\psi}_{pi}, \bar{\psi}_{ri}$ also belong to the same sector. Define

$$V(p,\dot{p}) = p^T C_s p + \dot{p}^T A_s \dot{p} + 2\sum_{i=1}^{3} \int_0^{w_i(t)} \psi_{ai}\{G_{pi}\bar{\psi}_{pi}(\nu)\}d\nu \tag{123}$$

where G_{pi} and w_i denote the ii^{th} and i^{th} elements of G_p and w respectively. This form is the well-known "Lure'-Postnikov type" Lyapunov function [Vid.78]. From Eqs. (29) and (31), $p^T C_s p + \dot{p}^T A_s \dot{p} = 0$ only when $\dot{\alpha} = 0$, $q = \dot{q} = 0$. That is, this quantity can be zero when $\alpha \neq 0$. However, when $q = 0, w = -\alpha$, which is nonzero when $\alpha \neq 0$. Thus the third term on the right hand side of (123) is positive (since ψ_{ai} is nonzero in a neighborhood of the origin) for $\alpha \neq 0$. Therefore, V is positive definite. From (123), using (27), (119), (89)-(91),

$$\begin{aligned}\dot{V} &= -2\dot{p}^T B_s \dot{p} - 2\sum_{i=1}^{3} w_i[\psi_{ai}\{G_{pi}\bar{\psi}_{pi}(w_i) + G_{ri}\bar{\psi}_{ri}(\dot{w}_i)\} \\ &\quad - \psi_{ai}\{G_{pi}\bar{\psi}_{pi}(w_i)\}]\end{aligned} \tag{124}$$

If ψ_{ai} are monotonic nondecreasing and ψ_{ri} belong to the $(0,\infty)$ sector, $\dot{V} \leq 0$, and the system is Lyapunov-stable.

If ψ_{ai} are monotonic increasing and ψ_{ri}, ψ_{pi} belong to the $(0,\infty)$ sector, $\dot{V} \leq -2\dot{q}^T D\dot{q}$, and $\dot{V} = 0$ only when $\dot{q} = 0$ and $\dot{w} = 0$, which implies $\dot{\alpha} = 0$. Considering the closed-loop equation,

$$\begin{bmatrix} 0 \\ \Lambda q \end{bmatrix} = \begin{bmatrix} I \\ \Phi^T \end{bmatrix} \psi_a\{-G_p\psi_p(y_p)\} \tag{125}$$

which yields $\psi_a\{-G_p\psi_p(y_p)\} = 0$ and $q = 0$. If ψ_{pi} belongs to the $(0,\infty)$ sector, $\psi_{ai}(\nu) = \psi_{pi}(\nu) = 0$ only when $\nu = 0$. Therefore, $\alpha = 0$. Thus $\dot{V} = 0$ only at the origin, and the system is globally asymptotically stable (g.a.s.). ■

Thus the CAC preserves stability in the presence of monotonic increasing actuator nonlinearities and sensor nonlinearities belonging to the $(0, \infty)$ sector (i.e., the sensor nonlinearities do not have to be monotonic increasing).

We next investigate the robustness properties of CDEC.

Robustness of CDEC

Stability With Dynamic Operator in the Loop- Consider the case where a nonlinear dynamic operator $\mathcal{H}(z_0; v)$ is present in the loop. Suppose $\mathcal{H}$ is represented by the following state-space model:

$$\dot{z}(t) = f(z, v, t); \quad z(0) = z_0 \tag{126}$$

$$w(t) = s(z, t) \tag{127}$$

where v and w are $\ell \times 1$ vectors that are the input and the output of $\mathcal{H}$. Define the operator

$$\partial\mathcal{H}(z_0; g) = \mathcal{H}(z_0; g) - \mathcal{H}(z_0; 0) \tag{128}$$

We define $\mathcal{H}$ to be internally stable if $\|\mathcal{H}(z_0; 0)\|$ is finite for any finite z_0.

Theorem 11. *Consider the system given by Eqs. (19), (20), (88) and (92), where the operator $\mathcal{H}$ has the state-space representation given by (126) and (127). Suppose $G_r^{-1}\partial\mathcal{H}$ is passive and $\mathcal{H}$ is uniformly observable, finite gain, internally stable, continuous operator. Then the closed-loop system is g.a.s.*

Proof. Defining

$$V(t) = q^T \Lambda q + \dot{q}^T \dot{q} \tag{129}$$

$V(t) \geq 0$ for all $t > 0$. Differentiating $V(t)$ with respect to t and using Eqs. (19), (20), (88) and (92),

$$\dot{V} = -2\dot{q}^T D\dot{q} - 2u_{cr}^T G_r^{-1} \mathcal{H}(z_0; u_{cr}) \tag{130}$$

Integrating from 0 to T, since $V \geq 0$,

$$0 \leq V(T) = V(0) - 2 < \dot{q}, D\dot{q} >_T -2 < u_{cr}, G_r^{-1}\mathcal{H}(z_0; u_{cr}) >_T \tag{131}$$

which yields (after manipulation)

$$2\lambda_m(D)\|\dot{q}\|_T^2 \leq V(0) - \beta + 2\|\dot{q}\|_T\|\Gamma\|_s\|\mathcal{H}(z_0; 0)\| \tag{132}$$

where β is a constant (see eq. 99). By using procedure similar to that in the proof of Theorem 1, it can be proved that $\|\dot{q}\|$ is bounded, and that the system is g.a.s. ■

The following corollaries are an immediate consequence of Theorem 11.

Corollary 11.1.. *If $\mathcal{H}$ is a asymptotically stable, observable, LTI operator with rational, mimimum-phase transfer matrix H(s), the closed-loop system of Eqs. (19), (20), (88) and (92) is a.s. if $H(s)G_r$ is positive real. That is,*

$$H(j\omega)G_r + G_r H^*(j\omega) \geq 0 \quad \text{for all real} \quad \omega. \tag{133}$$

■

Corollary 11.2. *Under the assumptions of Corollary 11.1, if G_r and H(s) are diagonal, the closed-loop system of Eqs. (19), (20), (88) and (92) is a.s. if*

$$\text{Re}[H_i(j\omega)] \geq 0 \quad \text{for all real} \quad \omega\,(i = 1, 2, ..., m) \tag{134}$$

■

As a result of Corollary 11.2, CDEC can tolerate stable first-order dynamics in the loop. If $H_i(s) = e^{-j\phi_i}$, we have $\text{Re}[H_i(j\omega)] \geq 0$ for $-90^o \leq \phi_i \leq 90^o$; therefore, CDEC have at least 90 degress phase margin.

Stability in the Presence of Nonlinearities- Suppose the actuators and sensors have nonlinearities ψ_a and ψ_r respectively, so that

$$u = \psi_a[-G_r \psi_r(\tilde{y}_r)] \tag{135}$$

where ψ_a and ψ_r are allowed to be time-varying. The following theorem gives sufficient conditions for global asymptotic stability.

Theorem 12. *Consider the closed-loop system given by Eqs. (19), (20) and (135), where G_r is diagonal with positive entries, and ψ_{ai}, ψ_{ri} belong to the $[0, \infty)$ sector. Then the closed-loop system is globally asymptotically stable.*

Proof. Starting with V as in Eq. (129) and defining $\bar{\psi}_{ai}(\nu) = -\psi_{ai}(-\nu)$,

$$\dot{V} = -2\dot{q}^T D\dot{q} - 2\sum_{i=1}^{m} \tilde{y}_{ri}\bar{\psi}_{ai}\{G_{ri}\psi_{ri}(\tilde{y}_{ri})\} \tag{136}$$

Thus $\dot{V} \leq 0$, and $\dot{V} \equiv 0$ only if $\dot{q} \equiv 0$, which can happen (from the equations of motion) only when $q \equiv 0$. Therefore, the system is globally asymptotically stable. ■

It should be noted that all the nonlinearities are allowed to be time- varying for this case. The next theorem considers a special case when nonlinearities and first-order dynamics are simultaneously present in the loop (e.g., representing the actuators) as shown in Figure 4.

Theorem 13. *Consider the closed-loop system given by Eqs. (19), (20), (88), and (92), where $G_r > 0$ is diagonal. Suppose $\mathcal{H} = diag(\mathcal{H}_1, \mathcal{H}_2, ..., \mathcal{H}_m)$, where*

$$\mathcal{H}_i g = \psi_i(\mathcal{G}_i g) \tag{137}$$

where $\psi_i : R \to R$ is a time-invariant, Lipschitz-continuous function belonging to the $[0, \infty)$ sector. Suppose $\mathcal{G}_i$ is an LTI operator whose transfer function is: $G_i(s) = a_i/(1 + p_i s), a_i > 0$, $p_i > 0$ for $i = 1, 2, .., m$. Then the system is g.a.s.

Proof. Starting with V as in Eq. (129) and proceeding as in the proof of Theorem 11, we have

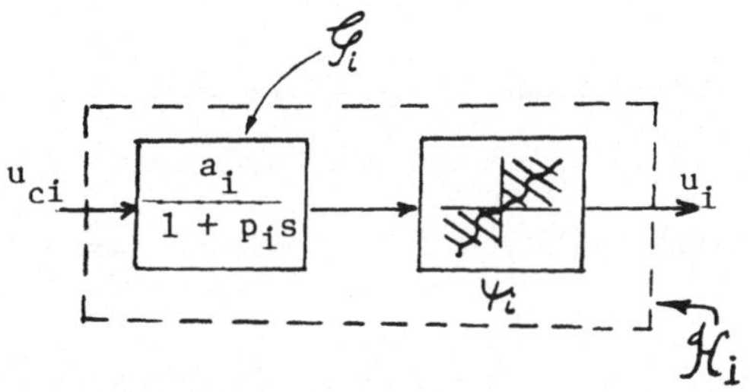

Figure 4. Linear dynamics and nonlinearities simultaneously in the loop

$$0 \leq V(0) - 2 < \dot{q}, D\dot{q} >_T -2 \sum_{i=1}^{m} G_{ri} < u_{cri}, \psi_i\{\mathcal{G}_i(0; u_{cri}) + g_{oi}\} >_T \tag{138}$$

where g_{oi} is the unforced response of $\mathcal{G}_i$ due to nonzero initial state. Eq. (138) can be written as:

$$\begin{aligned} 0 \leq V(0) - 2 < \dot{q}, D\dot{q} >_T -2 \sum_{i=1}^{m} < u_{cri}, \psi_i\{\mathcal{G}_i(0; u_{cri})\} >_T \\ + < u_{cri}, \psi_i\{\mathcal{G}_i(0; u_{cri}) + g_{oi}\} - \psi_i\{\mathcal{G}_i(0; u_{cri})\} >_T \end{aligned} \tag{139}$$

since each ψ_i is Lipschitz-continuous, there exists a constant K such that

$$|\psi_i(v_1) - \psi_i(v_2)| \leq K|v_1 - v_2| \quad i = 1, 2, ..., m$$

Noting that the operator $\psi_i\{\mathcal{G}_i(0; g)\}$ is passive [Des.75], and simplifying, we have

$$\lambda_m(D)\|\dot{q}\|_T^2 \leq V(0)/2 + \|\Gamma\|_s K \|\dot{q}\|_T \|g_o\| \tag{140}$$

where

$$\|g_o\| = \sum_{i=1}^{m} \|g_{oi}\| < \infty \tag{141}$$

The remainder of the proof is similar to that of Theorem 11. ■

From theorems 11-13, CDEC have excellent robustness to $[0, \infty)$ sector actuator and sensor nonlinearities, and also to actuator dynamics. These robustness properties are stronger than those of the CAC. We shall next investigate the robustness of the total velocity feedback controllers (TVFC) discussed in Sec. 2.2.

Robustness of TVFC

We shall prove that TVFC have robustness properties which are very similar to (but somewhat weaker than) those of CDEC. We shall consider the cases when the actuators/sensors have dynamics and nonlinearities. As discussed in Sec. 2.2, we omit the rigid-body attitude from the equations, and denote $\Omega = \dot{\alpha}$, so that the system given by:

$$[A_s]\begin{bmatrix}\dot{\Omega}\\ \ddot{q}\end{bmatrix} + [B_s]\begin{bmatrix}\Omega\\ \dot{q}\end{bmatrix} + \begin{bmatrix}0_{3\times 1}\\ \Lambda q\end{bmatrix} = \Gamma^T u \tag{142}$$

The TVFC command input is given by:

$$u_c = -G_r y_r \tag{143}$$

where G_r is symmetric and positive definite, and y_r is given by eq. (35).

Theorem 14. *Consider the closed-loop LFSS/TVFC system given by eqs. (35), (88), (142) and (143), where $\mathcal{H}$ is an asymptotically stable, observable LTI operator with a proper rational minimum phase transfer matrix H(s). The closed-loop system is* a.s. *if*

i) the closed-loop system with rigid model alone (i.e., eq. (142) with $n_q = 0$) is a.s., *and*

ii) $H(s)G_r$ is positive real.

Outline of Proof. The proof is very similar to that of Theorems 8 and 9 combined. The first step is to prove the asymptotic stability for the case: $H(s) = \varepsilon I + \hat{H}(s)$, $\varepsilon > 0$. This is done by starting with V as in eq. (41), and arriving at an inequality similar to (104). The Parseval theorem is then used to prove the asymptotic stability. A procedure similar to that of Theorem 9 is subsequently used to establish the asymptotic stability for $\varepsilon = 0$. ∎

The implication of this result is that for diagonal G_r and H the stability is guaranteed for $\text{Re}[H_i(j\omega)] \geq 0$ (equivalently, 90 deg. phase margin), which is the same as for the CDEC.

For the case where actuator and sensor nonlinearities ψ_a and ψ_r exist, considering the case where G_r is diagonal, the actual input is given by:

$$u = \psi_a\{-G_r\psi_r(y_r)\} \tag{144}$$

Starting with V as in (41) and following a procedure similar to that in the proof of Theorem 10, the following result can be obtained:

Theorem 15. *The closed-loop LFSS/TVFC system given by eqs. (35), (142) and (144) is g.a.s. if i) $G_r > 0$, diagonal, and ii) ψ_{ai}, ψ_{ri} belong to the $(0, \infty)$ sector.* ∎

In Theorem 15, the nonlinearities are allowed to be time-varying. Also note that for the TVFC case we require that the nonlinearities belong to the $(0, \infty)$ sector rather than the $[0, \infty)$ sector for the CDEC. Finally, for the case where the feedback loop consists of a first-order stable LTI dynamics followed by a nonlinearity, the following result can be obtained in a manner similar to the CDEC case.

Theorem 16. *Consider the closed-loop system given by (35), (88), (142) and (143), where $G_r > 0$ is diagonal, $\mathcal{H} = diag(\mathcal{H}_1, \mathcal{H}_2, ..., \mathcal{H}_m)$, where $\mathcal{H}_i$ is as defined in eq. (137). Suppose, for each i, $\psi_i : R \rightarrow R$ is a time invariant, Lipschitz-continuous single-valued function belonging to the $(0, \infty)$ sector, and $\mathcal{G}_i$ are as defined in the statement of Theorem 13. Then the closed-loop system is g.a.s.* ∎

The proofs of Theorems 14, 15 and 16 are basically similar to those of (respectively) Theorems 8 and 9; Theorem 12, and Theorem 13.

2.4.3 Summary of Robustness Results

Robustness properties were investigated for two types of dissipative controllers for large space structures. The first type is the collocated attitude controller (CAC), which controls the rigid-body attitude and the elastic motion using negative definite feedback of measured attitude and rate. The second type of controller employs velocity feedback for controlling mainly the elastic motion, and includes a) collocated damping enhancement controllers (CDEC), and b) total velocity feedback controllers (TVFC). In Sec. 2.2, such controllers were shown to provide closed-loop asymptotic stability regardless of the number of modes and parameter values, provided that the actuators

Table 1. Summary of robustness properties

Feedback operator	STABILITY CONDITIONS — CONTROLLER: Attitude (CAC) Stable if	Damping (CDEC) Stable if
I. LTI Dynamics		
a. general case	G_p, G_r, H satisfy (114)	$H(s)G_r$ positive real
b. G_p, G_r, H diagonal	$\mathrm{Arg}[H_i(j\omega)]$ satisfies (112)	$H_i(s)$ positive real
c. $H_i(s) = \frac{k_i}{s + a_i}$	$G_{pi} \leq a_i G_{ri}$	always stable
II. Nonlinearities in the		
a. Actuators	monotonic increasing, time-invariant (TI)	$[0,\infty)$ sector
b. Attitude sensors	$(0,\infty)$ sector, TI	....(not used)
c. Rate sensors	$(0,\infty)$ sector	....$[0,\infty)$ sector
III. Simultaneous TI nonlinearities and dynamics as in Fig. 4	No guaranteed stability	always stable

and sensors are perfect. In Sec. 2.4.2, this robust stability property was extended further by proving that the global asymptotic stability is preserved even when sensors and actuators are not perfect. The results are summarized in Table 1. In particular, the CAC preserves global asymptotic stablity when the sensors/actuators are represented by linear, time-invariant (LTI) dynamics which satisfy certain simple phase conditions. The attitude and rate sensors used in practice usually have high bandwidth. Thus a practical implication of this result is that one can always design a stable CAC if the actuators can be represented by first-order dynamics. The CAC was also shown to preserve stability when the actuators have monotonic nondecreasing nonlinearities, and the sensors have nonlinearities belonging to the $[0,\infty)$ sector. Many sensor and actuator nonlinearities encountered in practice (e.g., saturation, dead zone, etc.) have these characteristics. Thus a practical implication of this result is that the CAC will maintain stability when such commonly-encountered nonlinearities are present. The results obtained also indicate that the CAC can maintain stability in the presence of actuator and sensor outages, provided that at least one actuator and sensor per axis is functional.

The collocated damping enhancement controller (CDEC) was shown to preserve global asymptotic stability under much weaker conditions. In particular, the CDEC has 90 deg. phase margin and is tolerant to time-varying nonlinearities in the $[0,\infty)$ sector. The CDEC can also tolerate such nonlinearities and first-order actuator dynamics simultaneously in the loop. The total velocity feedback controller (TVFC) was shown to have essentially the same robustness properties as the CDEC, the only difference being that the nonlinearities are required to belong to the $(0,\infty)$ sector. (It can be shown that Lyapunov stability, rather than asymptotic stability, is preserved if the nonlinearities belong to the $[0,\infty)$ sector).

The stability results presented are valid regardless of the number of modes in the model and regardless of parameter values. Therefore, it can be concluded that dissipative controllers offer viable methods for robust attitude control or damping enhancement, especially when the parameters are not accurately known. An important application of dissipative controllers would be during *deployment* or *assembly* or *construction* of a large space structure, when the dynamic characteristics are changing, and during *initial operating phase*, when the dynamic characteristics are not known accurately. A robust CAC or TVFC can provide stable interim control which can perhaps be replaced later by a high-performance controller designed using parameters estimated on orbit.

2.5 Example: Dissipative Controller Design for a Large Space Antenna

In this subsection, we consider the design of an attitude control system for the 122 m. diameter hoop/column antenna described in Chapter 1. In order to achieve the required radio-frequency (RF) performance, the antenna must be controlled to specified precision in attitude and shape. For example, for missions such as the land mobile satellite system (LMSS), which is a communications concept for providing mobile telephone service to the continental United States, it is necessary to achieve a pointing accuracy of 0.03 degree root mean square (RMS) and a surface accuracy of 6 mm RMS [Jos.83a]. It is also necessary to have stringent control (usually a fraction of a degree) on the the motion of the feed (located near one end of the mast) relative to the mesh.

We consider only torque actuators located on the mast for controlling the antenna attitude and elastic motion. Because of the geometry of the antenna, reaction jets located on the hoop might be more effective in controlling the rigid-body roll mode and the torsion modes; however, because of their propellant storage requirements and the hardware difficulties in generating precise control forces, we do not include reaction jets for the actuation in this design. The surface accuracy is of extreme importance for the successful operation of the antenna. The surface can be actively controlled by pulling the control stringers; however, from practical considerations, it is preferable to avoid active surface control if at all possible, and to control the surface using only the torque actuators on the mast. As described in Chapter 1, the modes which cause surface distortion also cause bending and/or torsion of the mast, and are therefore controllable using the torque actuators located on the mast. Herein we assume four, three-axis torque actuators, and attitude and rate sensors, located on the mast at the locations shown in Figure 5.

The basic design objectives are: 1) to obtain sufficiently high closed-loop bandwidth for the rigid-body modes, and satisfactory damping ratios for both the rigid and elastic

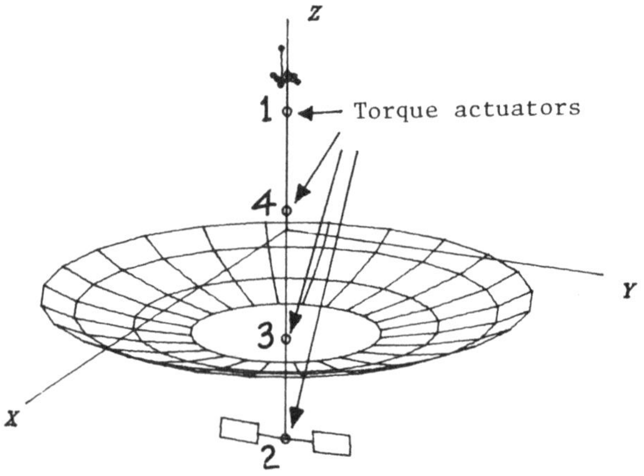

Figure 5. Torque actuator locations

modes, and 2) to obtain satisfactory RMS pointing errors, surface errors, and feed motion errors, in the presence of sensor and actuator noise. The first objective (i.e., bandwidth and damping ratios) arises from the need to obtain sufficiently rapid error decay at the end of a re-targeting maneuver, or when a step disturbance (such as sudden thermal distortion caused by entering or leaving Earth's shadow) occurs. The second design objective arises from the RF performance requirements. These two objectives may not necessarily be compatible, and may even be conflicting. For example, the use of higher feedback gains can yield higher bandwidth and damping ratios, but will generally result in increased RMS errors (because of the amplified effect of the sensor noise) beyond a certain point. Therefore, it is necessary to carefully consider the tradeoffs between the speed of response and the RMS errors.

The attitude controller design was accomplished using the approximate pole placement technique described in sec. 2.3. The first step in this procedure is to define the "desired" coefficient matrices B_d and C_d. Our aim was to perform a parametric study by varying the controller parameters and computing the resulting RMS performance. As a part of the first design objective, the desired rigid-body closed-loop bandwidth was varied in the range: 0.02- 0.25 rad/sec, and the desired real part of the closed-loop eigenvalues corresponding to the structural modes, $(\rho\omega)_d$, was varied from 0 to 0.5.(An inherent damping ratio of 0.01 was assumed for all the structural modes, and $(\rho\omega)_d = 0$ implies "no additional desired damping"). That is, the desired closed-loop eigenvalues corresponding to the structural modes are required to lie to the left of the vertical line through $-(\rho\omega)_d$ in the complex plane. The desired rigid-body damping ratio ρ_α was held at 0.7.

The maximum number of modes (eigenvalue pairs) which can be placed under the collocated framework is equal to the number of inputs, 12. Therefore, in addition to the three rigid rotational modes, we can "place" upto nine elastic modes. If we place less than nine elastic modes, we can minimize the norm of the feedback gains. Alternatively, we can include more than nine elastic modes in the design, and get an eigenvalue placement which is close to the "desired" in the least-squares sense. In any case, however, G_p, G_r are always positive definite, and the stability is always guaranteed.

Herein we chose to control the first ten elastic modes in addition to the three rigid modes, which makes the "design" model a 26th-order one.

The performance of the control system can be evaluated by computing the RMS values of various errors in the presence of sensor noise, actuator noise, and other disturbances. Disturbances such as gravity gradient torques and solar pressure are low-frequency, predictable, and repeatable; therefore, they can be compensated in an open-loop manner. However, sensor and actuator noise represent very significant sources of errors. We assume the attitude and rate sensors to have additive white measurement noise with standard deviation intensities of 0.488 arc-second and 0.031 arc-second/sec respectively [Jos.78a]. Since the data on actuator noise were not readily available, it was assumed to be zero for nominal performance computation. However, the actuator noise was included in the computation of the parametrized data (to be explained later). The sensor and actuator noises enter the closed-loop equations additively, and the final closed-loop equation is of the form:

$$\dot{x} = \bar{A}x + \bar{B}v \tag{145}$$

where x is the complete state vector (for the evaluation model), A is a strictly Hurwitz closed-loop matrix, and B is the noise input matrix. The noise vector v consists of all the noise terms; i.e., for the attitude sensors, the rate sensors, and the actuators. The covariance matrix of the closed-loop system evolves according to the equation:

$$\dot{\Sigma} = \bar{A}\Sigma + \Sigma\bar{A}^T + \bar{B}V\bar{B}^T \tag{146}$$

where $\Sigma(t) = \mathcal{E}[x(t)x^T(t)]$ is the state covariance matrix, and V is the covariance intensity matrix of $v(t)$. If $v(t)$ is a stationary process, $\Sigma(t)$ approaches a steady state value $\bar{\Sigma}$ as $t \to \infty$.

Five measures of pointing performance were considered:

a) maximum (taken over all the points on the mast) RMS pointing errors about the X, Y and Z axes, denoted ε_ϕ, ε_θ and ε_ψ and (all errors include contributions of the three rigid rotational modes and all 20 elastic modes)

b) maximum RMS feed motion error (maximum taken over seven points corresponding to feeds and feed panels, with error at each point being defined as the resultant of the X, Y, and Z directional motions of each point relative to the point on the mast where the mesh surface intersects the mast), and

c) maximum RMS surface error (maximum taken over the resultant displacements from nominal positions, of 96 points on the surface).

Figure 6 shows the nominal performance of the collocated attitude controller for different values of the closed-loop rigid- body frequency, $\omega_\alpha = 0.02$ rad/sec, 0.1 rad/sec, and 0.25 rad/sec. (the same ω_α for all three axes, with damping ratio $\rho_\alpha = 0.7$). It is apparent from Figure 6 that the RMS pointing errors $\varepsilon_\phi, \varepsilon_\theta, \varepsilon_\psi$ decrease as $(\rho\omega)_d$ is increased. However, as ω_α is increased, the RMS pointing errors first decrease, and then increase. The nominal RMS errors are well below the allowable limits. For example, for $\omega_\alpha = 0.1$ rad/sec and $(\rho\omega)_d = 0.25$, we have $\varepsilon_\phi = 0.62 \times 10^{-3}$ deg., $\varepsilon_\theta = 1.0 \times 10^{-5}$ deg., $\varepsilon_\psi = 0.55 \times 10^{-3}$ deg., $\varepsilon_f = 0.08mm$, $\varepsilon_s = 0.14mm$ ($\varepsilon_f, \varepsilon_s$ denote the RMS feed motion and surface errors respectively).

For effectively designing a control system, more generic data will be helpful. Since the covariance intensities (V_p, V_r and V_a) of the three noises (the attitude sensors, rate sensors, and actuators) enter the covariance equation linearly, we can parametrize the data by activating each noise one at a time, and computing the five RMS errors ($\varepsilon_\phi, \varepsilon_\theta, \varepsilon_\psi, \varepsilon_f, \varepsilon_s$). Figures 7, 8, and 9 show the coefficients $\delta_{pi}, \delta_{ri}, \delta_{ai}$ ($i = \phi, \theta, \psi, f, s$), which represent the appropriate error variance (denoted by subscript i), obtained by making the noise intensities V_p, V_r, and V_a equal to unity one at a time, while the other two are being held at zero. As a result, any of the five performance measures ε_i ($i = \phi, \theta, \psi, f, s$) can be computed from a given set of actual noise variances as follows:

$$\varepsilon_i = (\delta_{pi} V_p + \delta_{ri} V_r + \delta_{ai} V_a)^{1/2} \tag{147}$$

where the units of ε_i are degree for $i = \phi, \theta, \psi$ and mm for $i = f, s$. The units of the noise variances are (rad), (rad/sec), and (ft-lb) respectively. The coefficients in Figures 7-9 are plotted for three values of ω_α: 0.02, 0.1, and 0.25 rad/sec in order to consider

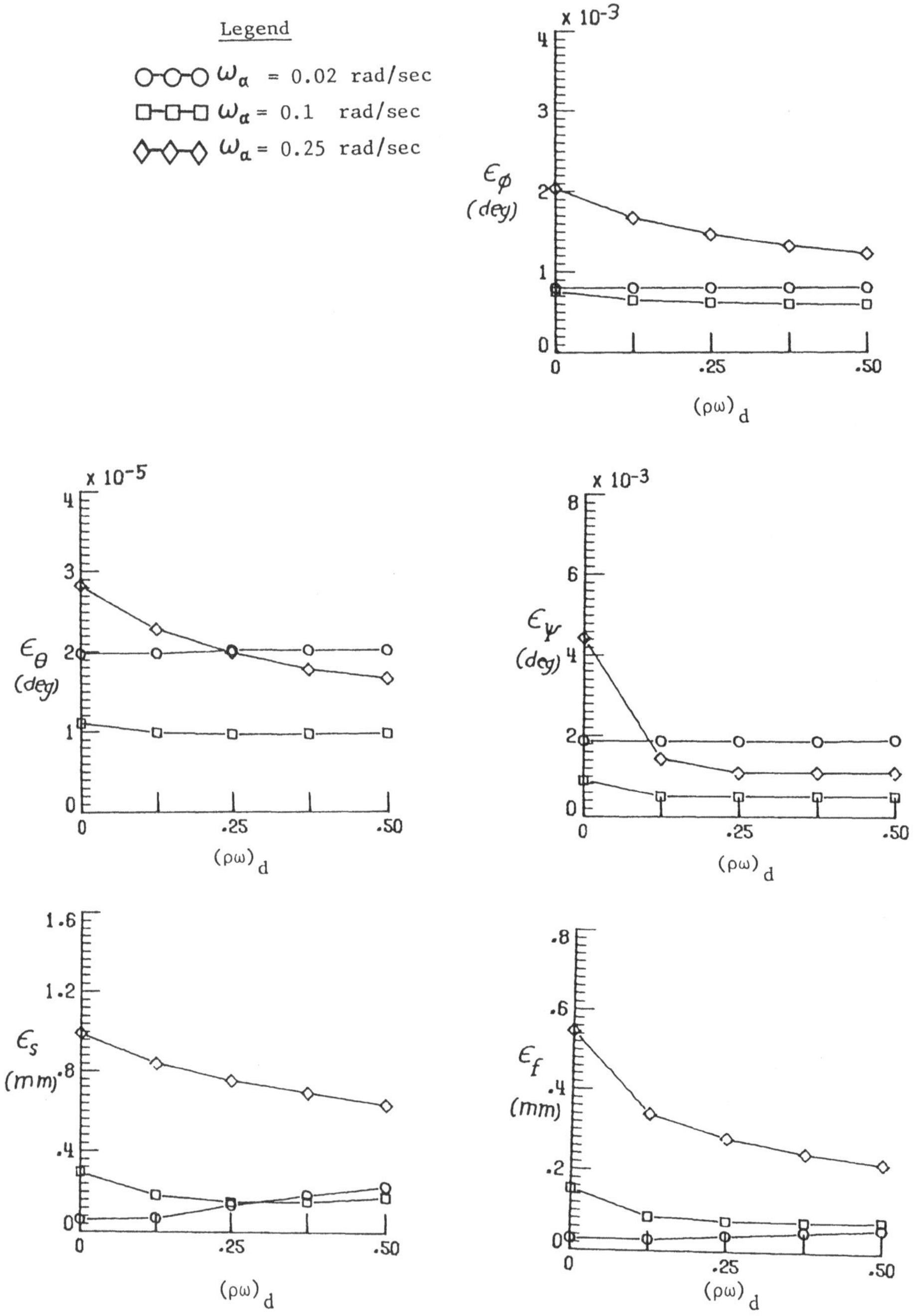

Figure 6. Performance of CAC

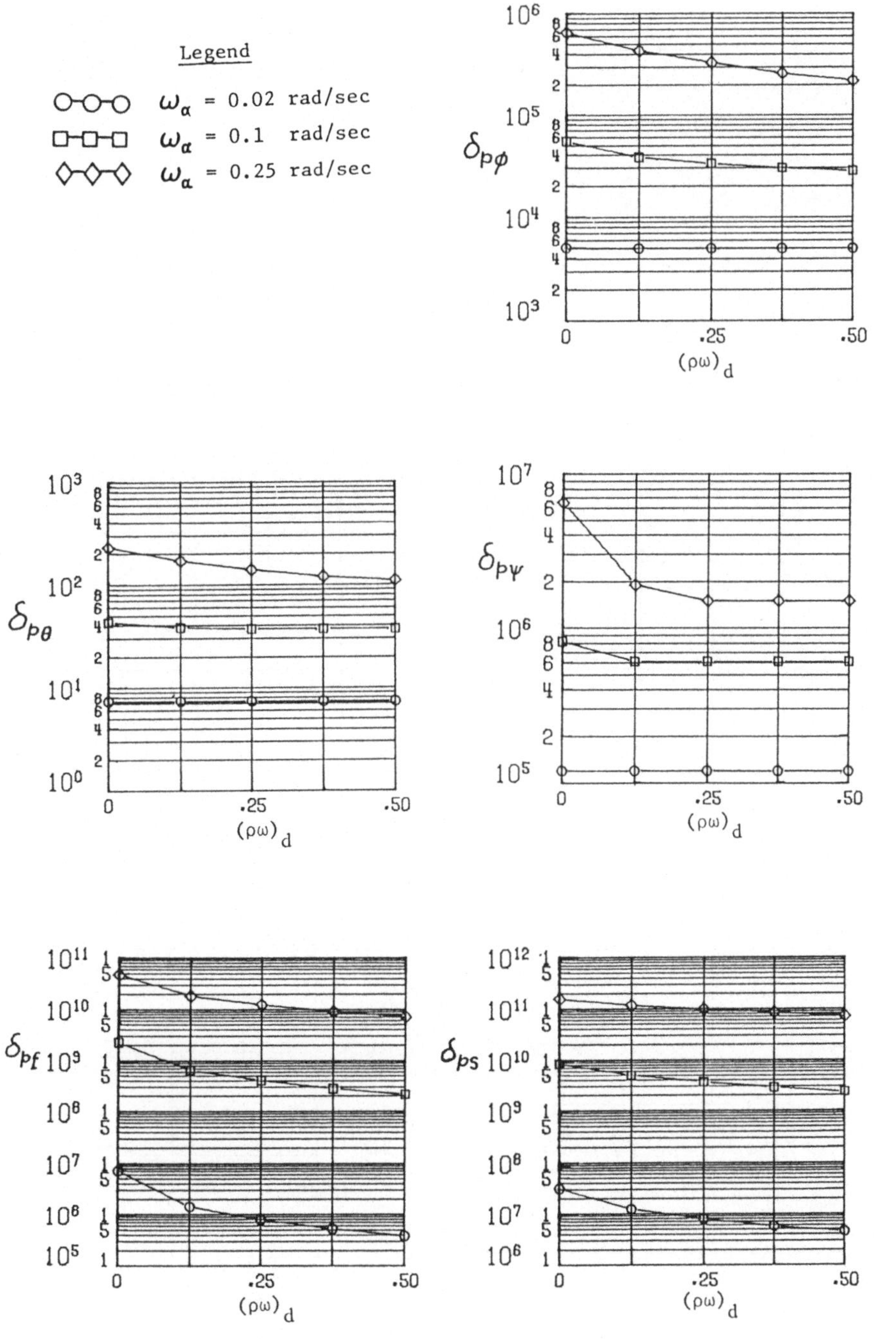

Figure 7. Performance coefficient δ_P

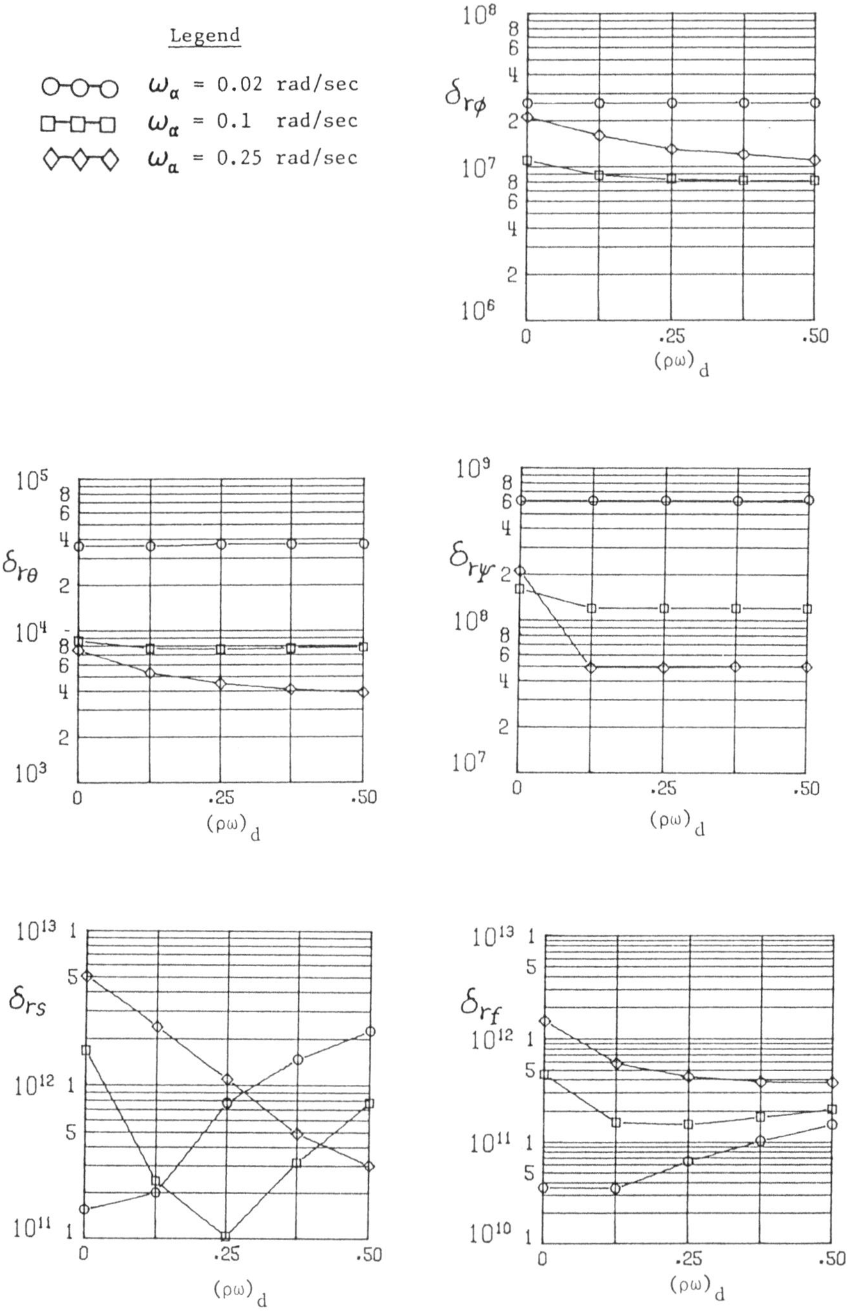

Figure 8. Performance coefficient δ_r

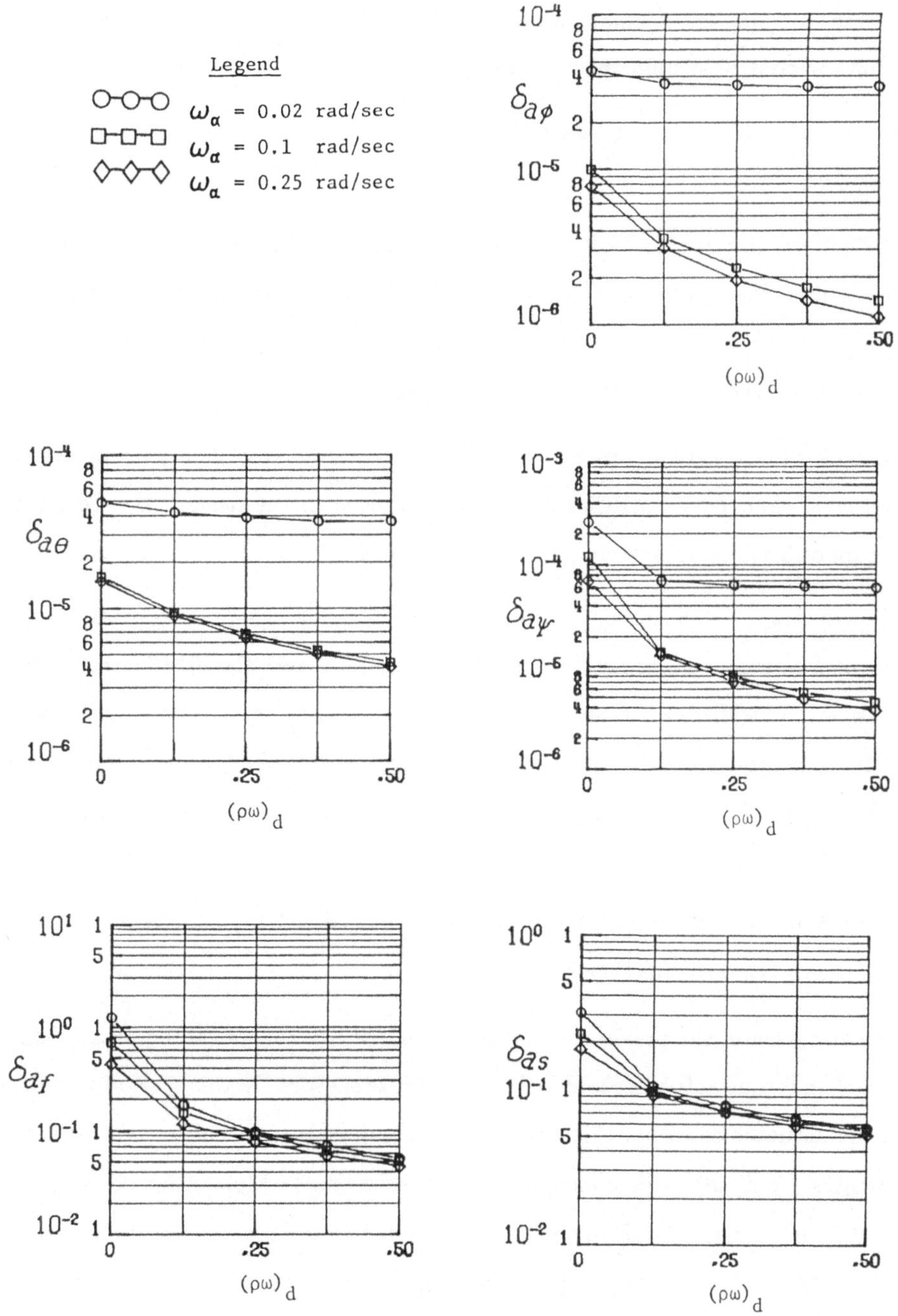

Figure 9. Performance coefficient δ_a

three response speeds. Generic data such as these can provide useful guidelines for selecting the control system specifications.

In order to evaluate the controller more completely, the following investigations were made:

Effect of imprecise collocation: All sensors were displaced from the corresponding actuator locations by ±60 cm along the mast. For the nominal case [$\omega_\alpha = 0.1$ rad/sec, $(\rho\omega)_d = 0.25$] the closed-loop eigenvalues remained virtually unchanged, and the RMS errors showed less than 1% increase.

Effect of imprecise knowledge of the parameters: The parameters (ω_i, ρ_i, and Φ) were changed by ±10% from the values used in the design; this resulted in a maximum of 2.5% deterioration in the performance.

In summary, a collocated attitude controller can be designed to meet the performance requirements in terms of both the bandwidth and the RMS error. For this antenna, a bandwidth of 0.1 rad/sec and $(\rho\omega)_d$ of 0.25 appear to be satisfactory. The controller performance was found to be relatively insensitive to parameter inaccuracies. A method of parametrizing the data was presented, which provides guidelines for selecting the design parameters.

2.6 The Annular Momentum Control Device (AMCD): An Actuator Concept for Dissipative Control

As discussed in the previous section, dissipative controllers can be realized in practice by using torque actuators and attitude and rate sensors. In this section we introduce a specific actuator concept which has the inherent feature of collocation of actuators and sensors. This actuator concept, called the "Annular Momentum Control Device (AMCD)", was originally developed at NASA-Langley Research Center [And.75] for attitude control of conventional (relatively rigid) spacecraft. However, because of its inherent design, it is also well suited for the actuation of dissipative controllers (with all the accompanying robustness properties), as will be shown in this section.

The conceptual development of the AMCD was motivated by the need to maximize the torque output to actuator weight ratio. The maximum output torque of a stored-

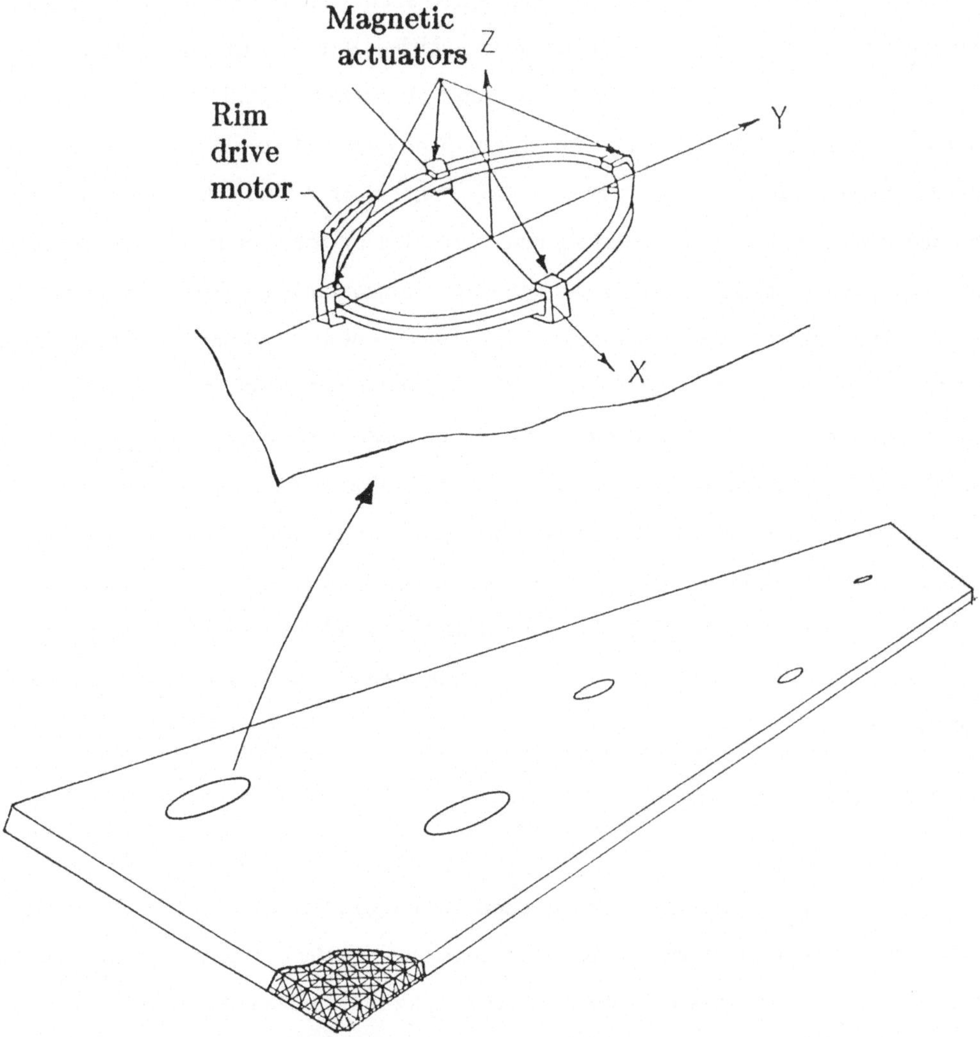

Figure 10. AMCD/LFSS configuration

momentum device (e.g., momentum wheels, CMG's, etc.) is proportional to the angular momentum which can be stored in its rotor, which is equal to the moment of inertia times the angular velocity. The optimum rotor shape for maximizing the moment of inertia for a given mass is an annular rim with no central hub, i.e., with all its mass concentrated at the rim. This is precisely the shape of an AMCD. As shown in Figure 10, the AMCD consists of a rotating rim suspended by three or more noncontacting electromagnetic actuator stations and spun by a noncontacting linear electromagnetic motor. Each electromagnetic actuator station consists of an axial and a radial actuator which can apply the commanded force in the corresponding direction. Axial and radial rim proximity sensors are also placed at the actuator locations; hence the inherent collocation property. The thickness of the rim is basically determined by the limit on the tip speed, which is proportional to the tensile stress. Increasing the rim radius increases the tip speed and the tensile stress. In addition, excessively large rim radius would make the rim flexible, thus complicating the problem. Therefore, we limit the AMCD rims in this analysis to about 2-3 meters diameter size, and assume that they are rigid.

2.6.1 Mathematical Model

Suppose a single AMCD is suspended using l (≥ 3) magnetic actuator stations. We are primarily concerned with rotations about two axes (X and Y, in the plane of the rim). The rotation about the Z-axis can be controlled by the spin motor (Z-axis torque can be applied simply by accelerating or decelerating the rim using the spin motor); and is not included in this analysis. The linearized two-axis AMCD rotational equations are given by: [Jos. 80a]

$$J_a \ddot{\alpha}_a + W \dot{\alpha}_a = C_1 f \tag{148}$$

where J_a is the transverse-axis rim inertia matrix (2×2), $\alpha_a = (\phi_a, \theta_a)^T$ are the rim rotation angles about the X and Y axes,

$$C_1 = \begin{bmatrix} y_1, & y_2, & \dots, & y_\ell \\ -x_1, & -x_2, & \dots, & -x_\ell \end{bmatrix} \tag{149}$$

(x_i, y_i) being the location of the ith actuator in the LFSS-fixed coordinate system with origin at the nominal position of the rim center.

$$W = \begin{bmatrix} 0 & H \\ -H & 0 \end{bmatrix} \tag{150}$$

where H is the rim angular momentum about the Z-axis. ($H = J_{az}\omega_z$, where ω_z is the rim spin velocity, which is assumed to be constant).

$$f = (F_1, F_2, ..., F_\ell)^T \tag{151}$$

where F_i is the axial (Z-direction) force at actuator i. The rim translation is given by:

$$m_a \ddot{z}_a = C_2 f \tag{152}$$

where m_a and z_a are the rim mass and the Z-axis displacement of the rim center, respectively, and

$$C_2 = (1, 1, ..., 1)_{1\times\ell} \tag{153}$$

The two-axis rigid-body equations of motion for the large, flexible space structure (LFSS) are given by:

$$J_s \ddot{\alpha}_s = -\xi C_2 f - C_1 f \tag{154}$$

where

$$\xi = (y_c, -x_c)^T \tag{155}$$

(x_c, y_c) being the location of the nominal position of the rim center in the LFSS-fixed coordinate system with origin at the LFSS c.m. J_s is the 2×2 LFSS inertia matrix, and $\alpha_s = (\phi_s, \theta_s)$ is the two-axis LFSS rigid-body attitude vector. The Z-axis rigid-body translational equation of the LFSS c.m. is given by:

$$m_s \ddot{z}_s = -C_2 f \tag{156}$$

where m_s is the LFSS mass and z_s is the Z-axis c.m. translation. The flexible motion of the LFSS is given by:

$$\ddot{q} + D\dot{q} + \Lambda q = -\Psi^T f \tag{157}$$

where q, D and Λ have been defined previously (Sec. 2.1), and Ψ is the $\ell \times n_q$ matrix of the Z-axis mode shapes at the actuator locations. As in the previous analysis, we assume that some damping is present in each flexible mode; i.e., $D = D^T > 0$.

Let z_{sa} denote the position of the location on the LFSS which corresponds to the nominal rim center position. The absolute translations z_{sa} and z_a are not important in this analysis; rather the relative translation ε is important, where

$$\varepsilon = z_a - z_{sa} = z_a - (z_s + \xi^T \alpha_s) \tag{158}$$

From eqs. (158), (152) and (156), we have

$$\ddot{\varepsilon} = m^{-1} C_2 f - \xi^T \ddot{\alpha} \tag{159}$$

where

$$m = m_a m_s / (m_a + m_s) \tag{160}$$

The equations of motion can be combined as:

$$A\ddot{x} + B\dot{x} + Cx = \beta f \tag{161}$$

where

$$x = (\alpha_a^T, \varepsilon, \alpha_s^T, q^T)^T \tag{162}$$

$$A = \begin{bmatrix} J_a & 0 & 0 & 0 \\ 0 & m & m\xi^T & 0 \\ 0 & m\xi & J_s + m\xi\xi^T & 0 \\ 0 & 0 & 0 & I_{n_q} \end{bmatrix} \tag{163}$$

$$B = \text{diag}(W, \quad 0_3, \quad D) \tag{164}$$

$$C = \text{diag}\left(0_5 \quad \Lambda_{n_q \times n_q}\right) \tag{165}$$

$$\beta = [C_1^T, \quad C_2^T, \quad -C_1^T, \quad -\Psi]^T \tag{166}$$

Using the transformation: $x = Th$, where

$$h = \left(\alpha_a^T, \quad \varepsilon, \quad \alpha_s^T - \alpha_a^T, \quad q^T\right)^T \tag{167}$$

$$T = \left[\begin{array}{ccc|ccc|c} & I_3 & & & 0 & & 0 \\ \hline 1 & 0 & 0 & 1 & 0 & & 0 \\ 0 & 1 & 0 & 0 & 1 & & 0 \\ \hline & 0 & & & 0 & & I_{n_q} \end{array}\right] \tag{168}$$

we have the equations in the form:

$$\hat{A}\ddot{h} + B\dot{h} + Ch = \gamma^T f \tag{169}$$

where

$$\hat{A} = T^T A T \tag{170}$$

$$\gamma = [0_{\ell\times 2}, C_2^T, -C_1^T, -\Psi] \tag{171}$$

The $\ell \times 1$ actuator (axial) centering error vector δ is given by:

$$\delta = C_1^T \alpha_s + \Psi q - C_1^T \alpha_a - C_2^T \varepsilon = -\gamma h \tag{172}$$

2.6.2 Damping Enhancement Using AMCD

We first consider the use of an AMCD for damping enhancement only, without attempting to control the attitude. In this mode, the control objective and the analysis are very similar to the total velocity feedback controller (TVFC) discussed in Sec. 2.1 and 2.2; that is, we are interested mainly in enhancing the structural mode damping and stabilizing the attitude (in the sense that attitude rate $\dot{\alpha}_s \to 0$).

Consider the magnetic actuator control law:

$$f = G_p \delta + G_r \dot{\delta} \tag{173}$$

where G_p and G_r are $\ell \times \ell$ symmetric positive definite matrices. Using eqs. (169), (172) and (173), the closed loop equation becomes:

$$\hat{A}\ddot{h} + \bar{B}\dot{h} + \bar{C}h = 0 \tag{174}$$

where

$$\bar{B} = B + \gamma^T G_r \gamma = \begin{bmatrix} W & 0 \\ 0 & \eta^T G_r \eta + \begin{pmatrix} 0 & 0 \\ 0 & D \end{pmatrix} \end{bmatrix} \tag{175}$$

$$\bar{C} = C + \gamma^T G_p \gamma = \begin{bmatrix} 0 & 0 \\ 0 & \eta^T G_p \eta + \begin{pmatrix} 0 & 0 \\ 0 & \Lambda \end{pmatrix} \end{bmatrix} \tag{176}$$

$$\eta = [C_2^T, -C_1^T, -\Psi] \tag{177}$$

In this configuration, the closed-loop system will always have two zero eigenvalues (corresponding to AMCD attitude α_a or equivalently, LFSS rigid-body attitude α_s), as shown below. Denoting

$$p = \left(\varepsilon, \alpha_a^T - \alpha_s^T, q^T\right)^T \tag{178}$$

Eq. (174) can be rewritten in the form:

$$\ddot{\alpha}_a = D_{11}\dot{\alpha}_a + D_{12}\dot{p} + E_{12}p \tag{179}$$

$$\ddot{p} = D_{21}\dot{\alpha}_a + D_{22}\dot{p} + E_{22}p \tag{180}$$

where D_{ij} and E_{ij} are appropriately dimensioned matrices. Since the right hand sides of eqs. (179) and (180) do not contain α_a, there are two free integrators in the system. Denoting $\dot{\alpha}_a = \omega_a$, we investigate the stability of the AMCD/LFSS system (174) after removing α_a, which is rewritten below as:

$$\frac{d}{dt}\begin{bmatrix} p \\ \omega_a \\ \dot{p} \end{bmatrix} = \left[\begin{array}{c|cc} 0_{(n_q+3)\times(n_q+5)} & & I_{n_q+3} \\ \hline E_{12} & D_{11} & D_{12} \\ E_{22} & D_{21} & D_{22} \end{array}\right] \begin{bmatrix} p \\ \omega_a \\ \dot{p} \end{bmatrix} \tag{181}$$

It should be noted that the removal of α_a from the closed-loop equations presents no problem because our main objective here is to achieve damping enhancement. The asymptotic stability of (181) would imply that, as $t \to \infty$, ω_a, $\varepsilon, (\alpha_a - \alpha_s)$, and q all tend to 0, and that α_a, α_s tend to constants (i.e., the LFSS would be stabilized). For controlling the attitude for normal operation, it would be necessary to employ an additional outer loop for attitude feedback. For now, we limit our attention to the asymptotic stability of the system (181).

In order to investigate the stability of eq. (181), we first prove the following lemmas:

Lemma 1. *For $\ell \geq 3$, the matrix L is always of full rank, where*

$$L = \begin{bmatrix} C_2 \\ -C_1 \end{bmatrix} = \begin{bmatrix} 1 & 1 & \dots & 1 \\ -y_1 & -y_2 & \dots & -y_\ell \\ x_1 & x_2 & \dots & x_\ell \end{bmatrix} \tag{182}$$

Proof. The first two columns of L (denoted by L_1 and L_2) are linearly independent. Suppose the ith column $L_i (i > 2)$ is linearly dependent on L_1 and L_2. Then there exist constants α_1 and α_2 such that

$$\alpha_1 L_1 + \alpha_2 L_2 = L_i \tag{183}$$

The top equation in (183) implies that $\alpha_1 + \alpha_2 = 1$. Therefore, from (182) and (183),

$$\alpha_1 L_1 + (1 - \alpha_1) L_2 = L_i \tag{184}$$

which implies that the actuator i is located on the straight line joining actuators 1 and 2. This is, of course, not true since the actuators are located on a circle. Thus the rank of L is 3. ■

Lemma 2. *If $G_p > 0$, the matrix $C_{22} > 0$, where*

$$C_{22} = \begin{bmatrix} L \\ -\Psi^T \end{bmatrix} G_p [L^T \quad -\Psi] + \begin{bmatrix} 0 & 0 \\ 0 & \Lambda \end{bmatrix} \tag{185}$$

Proof. Let $y = (y_1^T, y_2^T)^T$, where y_1 and y_2 are real 3×1 and $n_q \times 1$ vectors. Because $\Lambda > 0$, $y^T C_{22} y$ can be zero only if $y_2 = 0$, and $y_1^T L G_p L^T y_1 = 0$; i.e., only if $y_1 = 0$ (since rank of L is 3). ■

Theorem 17. *The closed-loop AMCD/LFSS system given by Eq. (181) is Lyapunov-stable if $G_p > 0$ and $G_r \geq 0$. The system is asymptotically stable if $G_p > 0$, $G_r > 0$, and $H \neq 0$.*

Proof. Consider the Lyapunov function:

$$V(p, \omega_a, \dot{p}) = p^T C_{22} p + \dot{h}^T \hat{A} \dot{h} \tag{186}$$

where $h = (\alpha_a^T, p^T)^T$. Since $A > 0$, $\hat{A}$ is also > 0. C_{22} was proved to be > 0 in Lemma 2. Therefore, $V > 0$. Differentiating V with respect to t and using (174), we get the following after simplification:

$$\begin{aligned} \dot{V} &= -\dot{h}^T (\bar{B} + \bar{B}^T) \dot{h} \\ &= -2(\dot{p}^T \eta^T G_r \eta \dot{p} + \dot{q}^T D \dot{q}) \end{aligned} \tag{187}$$

In writing the above equation we have utilized the fact that W is skew-symmetric for any H. If $G_r \geq 0$, this implies $\dot{V} \leq 0$, and the system is Lyapunov-stable.

Now suppose $G_r > 0$ and $H \neq 0$. Then $\dot{V}$ can be 0 only if $\dot{q} = 0$ and $\dot{p} = 0$ (since rank of L is 3), which implies p = constant. From eq. (174), this implies that $\dot{V} = 0$ only if

$$\hat{A}_{11}\dot{\omega}_a + W\omega_a = 0 \tag{188}$$

$$\hat{A}_{21}\dot{\omega}_a + C_{22}d_1 = 0 \tag{189}$$

where d_1 is a 2×1 constant vector, and $\hat{A}_{ij}$ are appropriate submatrices of $\hat{A}$, $\hat{A}_{11}$ being 2×2. Using Eq. (170), it can be verified that $\hat{A}_{21}^T = [m\xi, J_s + m\xi\xi^T, 0]$, which is of full rank. Therefore, $\dot{V} = 0$ only if $\dot{\omega}_a = d_2$, a constant, or if

$$\omega_a = d_3 + d_2 t \quad (d_3 = \text{constant})$$

i.e., only if

$$\hat{A}_{11}d_2 + W(d_3 + d_2 t) = 0$$

Since $H \neq 0$, W is nonsingular, and V can be zero only when $d_2 = d_3 = 0$. Thus $\dot{V} \neq 0$ along all trajectories, and the system is a.s. ■

2.6.3 Damping Enhancement Using Several AMCD's

In this subsection, we extend the mathematical model and stability analysis presented above, to the case where many AMCD's are used. As in the previous subsection, we assume the LFSS to be a large, flat, platform-type structure, with the spin axis of each AMCD being perpendicular to the plane of the structure (Fig. 10). Thus, only X and Y axis rotations, and Z-axis translations are considered in this configuration, which suffices to present the principles. It would be possible to extend this analysis in a straightforward but cumbersome manner, to general LFSS with the AMCD spin axes pointing in different directions.

We assume that ν AMCD's are distributed throughout the structure, and that the rim mass and the (2×2) transverse- axis rim intertia matrix of the ith AMCD are denoted by m_{ai} and J_{ai} respectively. Assuming that the ith AMCD uses ℓ_i actuator stations ($\ell_i \geq 3$), let C_{1i} and C_{2i} denote the $2 \times \ell_i$ and $1 \times \ell_i$ matrices defined similar to those for the single AMCD case (Eqs. 149 and 153). Let α_{ai} and ε_i denote the $2 \times \ell_i$ rim attitude vector and the Z-axis rim center displacement of the ith AMCD, and let f_i denote the $\ell_i \times 1$ axial actuator force vector. The equations of motion for the multiple AMCD case can then be derived in the form:

$$A\ddot{x} + B\dot{x} + Cx = \gamma^T f \tag{190}$$

where

$$x = (\alpha_s^T, \alpha_{a1}^T - \alpha_s^T, \alpha_{a2}^T - \alpha_s^T, ..., \alpha_{a\nu}^T - \alpha_s^T, \varepsilon_1, \varepsilon_2, ..., \varepsilon_\nu, q^T)^T \tag{191}$$

$$f = (f_1^T, f_2^T, ..., f_\nu^T)^T \tag{192}$$

$$A = \text{diag}\left[A_{1_{(3\nu+2)\times(3\nu+2)}}, I_{n_q}\right] \tag{193}$$

where

$$A_1^{-1} = \left[\begin{array}{c|cccc|c} J_s^{-1} & -J_s^{-1} & \cdots & & -J_s^{-1} & -J_s^{-1}\xi \\ \hline -J_s^{-1} & (J_s^{-1} + J_{a_1}^{-1}) & J_s^{-1} & \cdots & J_s^{-1} & J_s^{-1}\xi \\ \vdots & \vdots & & \ddots & \vdots & \vdots \\ -J_s^{-1} & J_s^{-1} \cdots & & \cdots & (J_s^{-1} + J_{a\nu}^{-1}) & J_s^{-1}\xi \\ \hline -\xi^T J_s^{-1} & \xi^T J_s^{-1} & & \xi^T J_s^{-1} \cdots & \xi^T J_s^{-1} & (\xi^T J_s^{-1}\xi + m_s^{-1} I_\nu + M_a^{-1}) \end{array}\right] \tag{194}$$

In (194)

$$\xi = (\xi_1, \xi_2, ..., \xi_\nu), \quad \xi_i = (y_i, -x_i)^T, \tag{195}$$

$$M_a = \text{diag}\,(m_{a1}, m_{a2}, ..., m_{a\nu}) \tag{196}$$

$$B = \left[\begin{array}{c|ccccc|c|c} \sum_{i=1}^{\nu} W_i & W_1 & W_2 & . & . & W_\nu & 0 & 0 \\ \hline W_1 & W_1 & & & 0 & & 0 & 0 \\ W_2 & & W_2 & & & 0 & 0 & 0 \\ & & & . & & & & \\ . & & & & . & & & \\ W_\nu & & & & & W_\nu & & \\ \hline 0 & & & & 0 & & 0 & 0 \\ \hline 0 & & & & 0 & & 0 & D \end{array}\right] \tag{197}$$

where

$$W_i = \begin{bmatrix} 0 & H_i \\ -H_i & 0 \end{bmatrix} \tag{198}$$

H_i being the Z-axis angular momentum of the ith AMCD rim.

$$C = \text{diag}[0_{(3\nu+2)}, \Lambda_{n_q \times n_q}] \tag{199}$$

$$\gamma^T = \begin{bmatrix} 0_{2\times\ell} \\ \text{diag}(C_{11}, C_{12}, ..., C_{1\nu}) \\ \text{diag}(C_{21}, C_{22}, ..., C_{2\nu}) \\ -\Psi_1^T, -\Psi_2^T, ..., -\Psi_\nu^T \end{bmatrix}_{n_2 \times \ell} \triangleq \begin{bmatrix} 0_{2\times\ell} \\ \eta_{n_1\times\ell} \end{bmatrix} \tag{200}$$

$\Psi_i (i = 1, 2, ..., \nu)$ represents $\ell_i \times n_q$ mode shape matrix for actuator locations of the ith AMCD, $\ell = \Sigma_{i=1}^{\nu} \ell_i$, $n_1 = n_q + 3\nu$, and $n_2 = n_1 + 2$. Let δ_{ik} denote the axial rim centering error at actuator station k of the ith AMCD. As in the single AMCD case, the rank of $[C_{1i}^T, C_{2i}^T]$ is 3, and the rank of γ is then 3ν. The $\ell \times 1$ rim centering error vector is given by:

$$\delta = [\delta_{11}, ..., \delta_{1\ell_1}, ..., \delta_{\nu 1}, ..., \delta_{\nu \ell_\nu}]^T = -\gamma x \tag{201}$$

As in the single AMCD case, consider the control law:

$$f = G_p \delta + G_r \dot{\delta} \tag{202}$$

where G_p and G_r are $\ell \times \ell$ real symmetric matrices.

The mathematical model presented above for the multiple AMCD case is derived in a manner very similar to the single AMCD case. The only difference is that the x vector is selected in a slightly different manner for convenience. It is obtained from the original vector (consisting of $\alpha_a, \varepsilon, \alpha_s$ and q) by using an appropriate similarity transformation. The closed-loop equations become:

$$A\ddot{x} + \bar{B}\dot{x} + \bar{C}x = 0 \tag{203}$$

where

$$\bar{B} = B + \gamma^T G_r \gamma \tag{204}$$

$$\begin{aligned} \bar{C} &= C + \gamma^T G_p \gamma \\ &= \text{diag } \left[0_2, C_{22_{(3\nu+2)\times(3\nu+2)}} \right] \end{aligned} \tag{205}$$

where

$$C_{22} = \eta^T G_p \eta + \text{diag}[0_{n_2}, \Lambda] \tag{206}$$

As in the previous case, there are always two zero eigenvalues corresponding to the two rigid LFSS rotational modes. Defining

$$p = \left(\alpha_{a1}^T - \alpha_s^T, ..., \alpha_{a\nu}^T - \alpha_s^T, \varepsilon_1, ..., \varepsilon_\nu, q^T\right)^T \tag{207}$$

we examine the stability of the closed-loop system consisting of the state vector: $(p^T, \dot{x}^T)^T$, which has the form:

$$\frac{d}{dt}\begin{bmatrix} \dot{p} \\ \dot{x} \end{bmatrix} = \frac{d}{dt}\begin{bmatrix} p \\ \dot{\alpha}_s \\ \dot{p} \end{bmatrix} = \begin{bmatrix} 0 & 0 & I_{n_1} \\ E_{1_{n_2 \times n_1}} & E_{2_{n_2 \times n_2}} & \end{bmatrix} \tag{208}$$

The first step is to prove that C_{22} given by eq. (206) is positive definite. This can be accomplished in a manner similar to the single AMCD case. The following theorem generalizes the stability result to the multiple AMCD case.

Theorem 18. *The closed-loop multiple AMCD/LFSS system given by Eq. (208) is stable in the sense of Lyapunov if $G_p > 0$ and $G_r \geq 0$. The system is asymptotically stable if $G_p > 0$, $G_r > 0$, and $\sum_{i=1}^{\nu} H_i \neq 0$.*

The proof is entirely analogous to that of Theorem 17, and will therefore be omitted.

The stability of the AMCD/LFSS sytem is guaranteed regardless of the model order or the lack of knowledge of the parameters. It should be noted that the system would still be Lyapunov stable for $\Sigma H_i = 0$; however, the damping enhancement, which will only be due to the rim masses, will be minimal.

2.6.4 Numerical Example

For the purpose of demonstration of the effectiveness of the AMCD for damping enhancement, we consider the 44-flexible mode finite element model of a $30.48m \times 30.48m \times 2.54mm (100ft. \times 100ft. \times 0.1in.)$ flexible aluminum plate, which was described in Chapter 1. The inherent damping was assumed to be zero for all the modes. Let us first investigate the effect of a single AMCD, with rim diameter of 1.79 m (5.88 ft.), rim mass 34 Kg (weight 75 lb.), and with four actuator stations, located on the plate as shown in Figure 11 as "AMCD No. 1". The nominal spin speed of the AMCD is assumed to be 5000 RPM. These parameters are chosen to be within the limits of the present day technology. Letting $G_p = g_p I$ and $G_r = g_r I$, g_p was fixed at 146 N/m, and

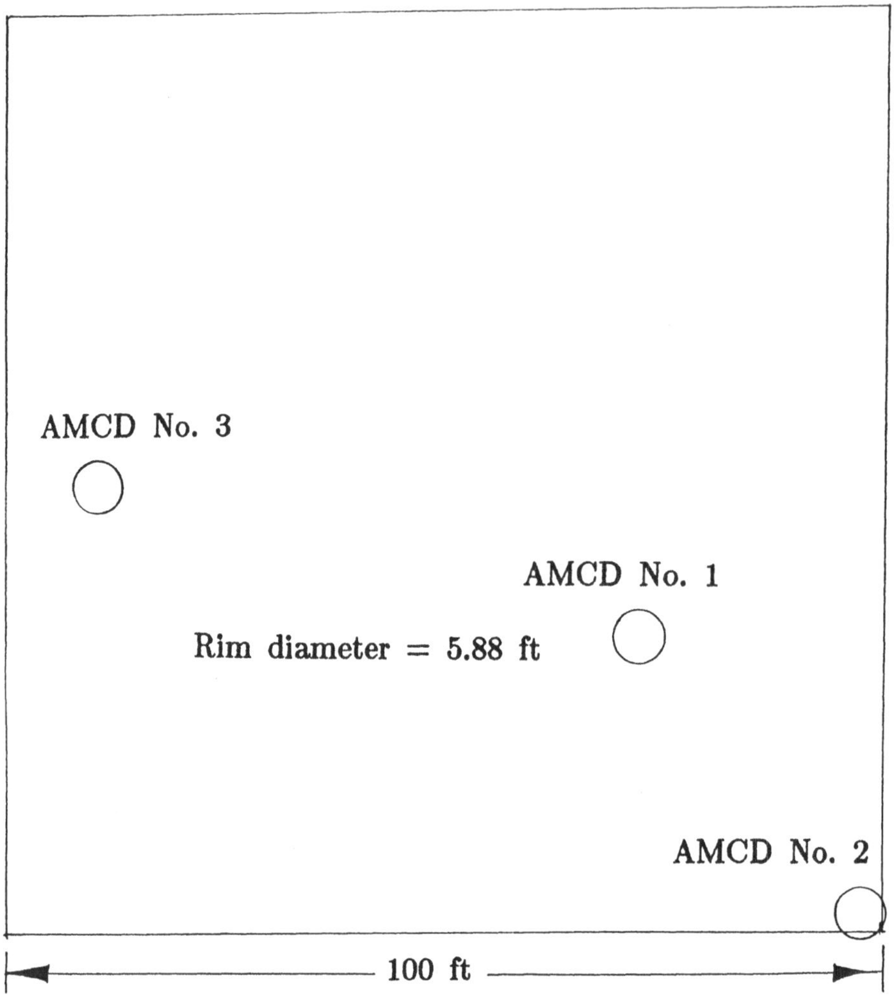

Figure 11. AMCD locations on LFSS

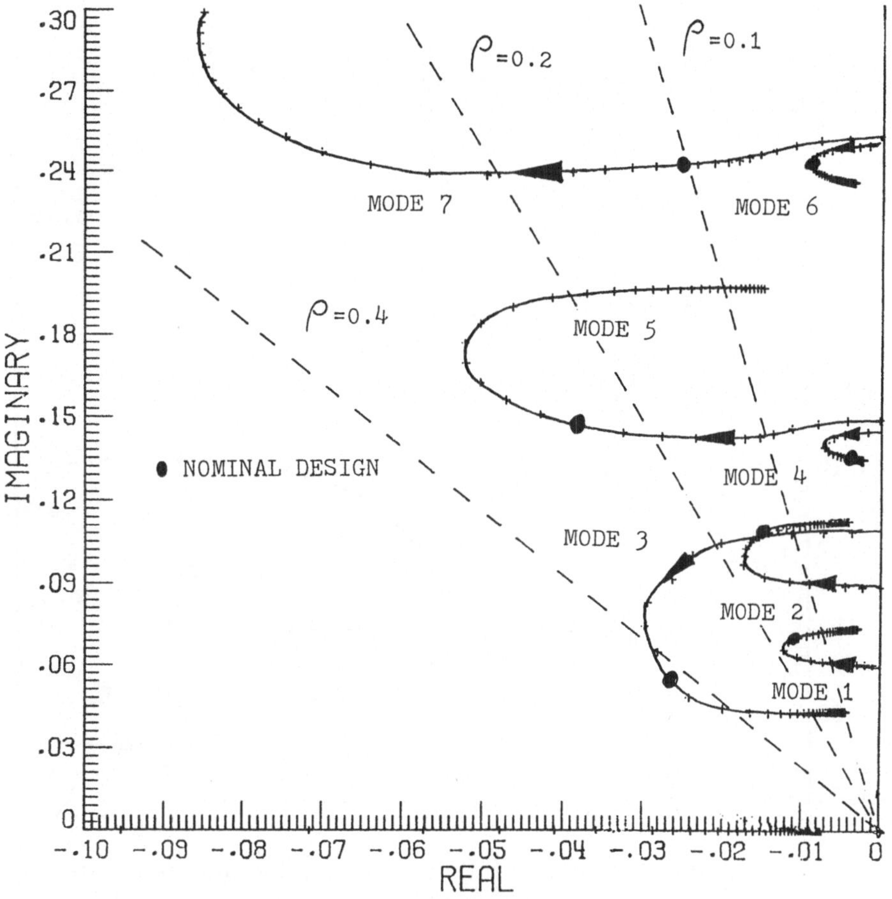

Figure 12. Root locus with respect to g_r

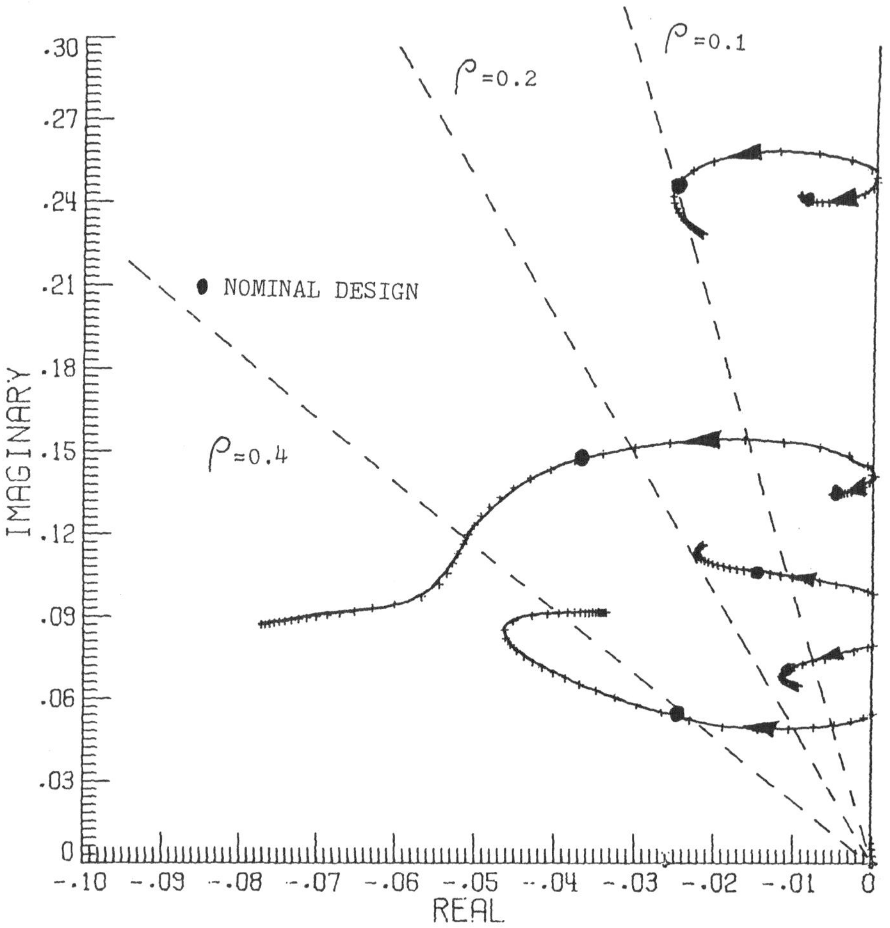

Figure 13. Root locus with respect to H

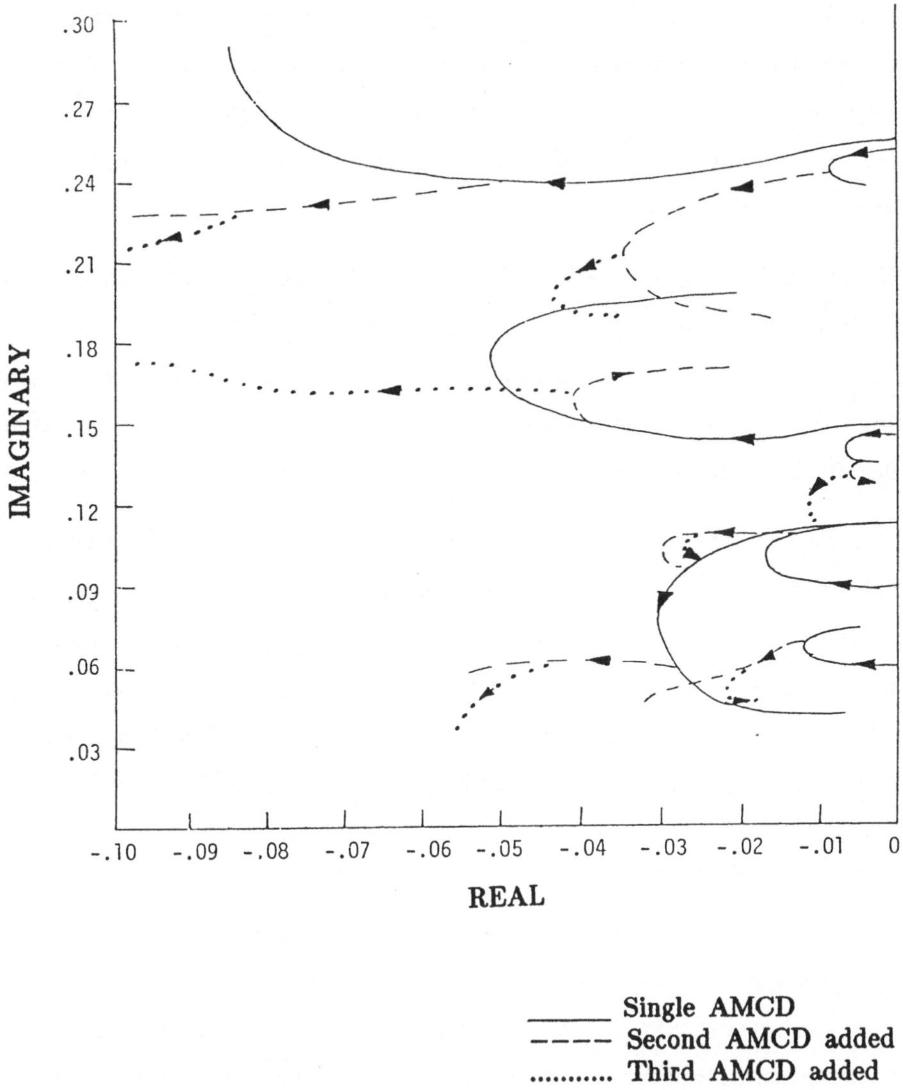

Figure 14. Root loci with three AMCD's

g_r was varied from 0 to 20,000 $N - sec/m$. The resulting root loci are shown in Figure 12 for the first seven modes. Damping for modes 3, 5, and 7 improved significantly, but that for modes 1, 4 and 6 improved only marginally, because the actuator location was more favorable to the former set of modes. In practice, the locations must be chosen to maximize the effect on the desired modes.

In order to investigate the effect of variation of the momentum, H was next varied from 0 to four times its nominal value, with the gains fixed at $g_p = 146N/m$ and $g_r = 5614N - sec/m$. At zero momentum, there is a small damping because the mass of the rim acts like a proof-mass actuator. As shown in Figure 13, the damping improved significantly for modes 1, 3, 5 and 7.

In order to study the effect of many AMCD's, two more identical AMCD's (No. 2 and No. 3) were next added at different locations as shown in Figure 11. Keeping the position gains of the AMCD's at: $g_{p1} = 146N/m$ and $g_{p2} = g_{p3} = 14.6N/m$, the rate gains were increased (starting from 0) for the first, then the second, and finally the third AMCD. The resulting root loci are shown in Figure 14. The figure shows that substantial damping enhancement is possible using multiple AMCD's.

2.6.5 Attitude Control for AMCD-Actuated LFSS

So far we have discussed damping enhancement using AMCD's. We have proved that the asymptotic stability of the system cosisting of $\dot{\alpha}_s, (\alpha_{ai} - \alpha_s), \varepsilon_i (i = 1, 2, ..., \nu)$, and q, is guaranteed; that is, these variables will tend to zero as $t \rightarrow \infty$. In other words, the structure will tend to a steady state as $t \rightarrow \infty$, although its attitude will tend to some (non-zero) constant. This property (which was also discussed in Sec. 2.2 in connection with TVFC) is extremely useful for stabilizing a structure which is vibrating or beginning to tumble. For example, such a controller would be very useful for maintaining the stability and integrity of the LFSS during deployment, assembly, or initial post-assembly phase, when the parameters are yet to be estimated, and a high-performance controller is yet to be designed.

During actual operation, however, in addition to damping enhancement, the attitude of the LFSS must be controlled. This can be accomplished in two ways: 1) using additional torque actuators and attitude and rate sensors, or 2) using the same AMCD's. We call the first method a "two level" controller, consisting of a secondary controller for damping enhancement, and a primary controller for attitude control.

Attitude Control Using a Two-Level Controller

With the secondary controller consisting of one or more AMCD's, suppose there are m_T, 2-axis torque actuators ($m_T \geq 1$), collocated with m_T, 2-axis attitude sensors and rate sensors. We begin with equation (203), wherein the secondary control loop is already closed. Because of the additional torque actuators, (203) is modified as:

$$A\ddot{x} + \bar{B}\dot{x} + \bar{C}x = \Gamma^T u \tag{209}$$

where $A, \bar{B}$ and $\bar{C}$ are defined in (193), (204), and (205), and

$$u = \left(T_1^T, T_2^T, ..., T_{mT}^T\right)^T \tag{210}$$

$$\Gamma^T = \begin{bmatrix} I_2 & \cdots & I_2 \\ & 0_{3\nu \times 2m_T} & \\ \Phi_1^T & \cdots & \Phi_{mT}^T \end{bmatrix} \tag{211}$$

where Φ_i represents the $2 \times n_q$ mode-slope matrix at the ith torque actuator location. The attitude and rate vectors at the actuator locations are given by (ignoring noise):

$$y_p = \Gamma x \tag{212}$$

and

$$y_r = \Gamma \dot{x} \tag{213}$$

Equations (209) - (213) have the same form as eqs. (4), (14), and (15). The main difference is that $\bar{B}$ in (209) is not symmetric. The top $(2\nu + 2) \times (2\nu + 2)$ principal submatrix of B (Eq. 197)is skew-symmetric. Consider the control law:

$$u = -\bar{G}_p y_p - \bar{G}_r y_r \tag{214}$$

We now investigate the stability of the entire system using the two-level controller.

Theorem 19. *The closed-loop system given by eqs. (209)- (214) is a.s. if $\bar{G}_p = \bar{G}_p^T > 0$ and $\bar{G}_r = \bar{G}_r^T > 0$.*

Proof. With the control law (214), (209) becomes:

$$A\ddot{x} + \hat{B}\dot{x} + \hat{C}x = 0 \tag{215}$$

where

$$\hat{B} = \bar{B} + \Gamma^T \bar{G}_r \Gamma \tag{216}$$

$$\hat{C} = \bar{C} + \Gamma^T \bar{G}_p \Gamma \tag{217}$$

We first show that $\hat{C}$ is positive definite if $\bar{G}_p > 0$. Indeed,

$$\bar{C} = \begin{bmatrix} 0 & 0 \\ 0 & C_{22} \end{bmatrix} \tag{218}$$

where C_{22} was previously proved to be positive definite. Since the rank of Γ is at least 2, it follows from (217) that $\hat{C}$ is positive difinite.

Consider the Lyapunov function

$$V(x, \dot{x}) = x^T \hat{C} x + \dot{x}^T A \dot{x} \tag{219}$$

As in the previous cases, we have

$$\dot{V} = -\dot{x}^T(\hat{B} + \hat{B}^T)\dot{x} = -\dot{x}^T(\bar{B} + \bar{B}^T + 2\Gamma^T \bar{G}_r \Gamma)\dot{x} \tag{220}$$

Using the definition of $\bar{B}$ from (204) and the skew-symmetry of W_i ($i = 1, 2, ..., \nu$),

$$\dot{V} = -2\dot{q}^T D\dot{q} - 2\dot{x}^T \left(\gamma^T G_r \gamma + \Gamma^T \bar{G}_r \Gamma\right) \dot{x} \tag{221}$$

where γ is defined in (200).

Since $D > 0, \dot{V} = 0$ implies $\dot{q} = 0$, and

$$\dot{\bar{p}}^T \eta^T G_r \eta \dot{\bar{p}} = \dot{\alpha}_s^T I^T \bar{G}_r I \dot{\alpha}_s = 0 \tag{222}$$

where

$$\bar{p} = \left(\alpha_{a1}^T - \alpha_s^T, ..., \alpha_{a\nu}^T - \alpha_s^T, \varepsilon_1, \varepsilon_2, ..., \varepsilon_\nu\right)^T \tag{223}$$

$$I = (I_2, I_2, ..., I_2)^T_{2\times 2m_T} \tag{224}$$

and η is defined in (200). Since η and I are of full rank, $\dot{V} = 0$ implies $\dot{\bar{p}} = 0$ and $\dot{\alpha}_s = 0$; i.e., $\dot{x} = 0$. The fact that $\hat{C} > 0$ then implies [(from 215)] that $\dot{V} = 0$ only if $x = 0$, and the system is a.s. ■

It was proved previously (Sec. 2.2) that the closed-loop asymptotic stability can be obtained with only torque actuators, without using AMCD's. However, the AMCD's significantly enhance the structural damping with relatively small weight penalty. One can accomplish the same damping enhancement using torque actuators; but they would add considerably to the weight, and therefore, to the mission cost. Because of their optimized shape, AMCD's are ideally suited for damping enhancement.

We next investigate the problem of attitude control using the same AMCD's; i.e., without using additional torque actuators.

Attitude Control Using AMCD's

In the fine-pointing mode, it is possible to control the attitude using the same AMCD's which are simultaneously being used for damping enhancement. This is accomplished by torquing against the AMCD angular momenta. However, in this dual control mode, we cannot simultaneously control the relative rotational angles $(\alpha_{ai} - \alpha_s)$

and rigid-body attitude α_s; the system consisting of both of these in its state vector can be easily shown to be uncontrollable. The dual control is accomplished in two steps, as follows: i) use the magnetic actuator control law (202) only for elastic motion damping and rim centering, and ii) superimpose the LFSS attitude control signal on the magnetic actuators based on the required attitude control torque,i.e., as in (214)]. In step (ii), we assume that attitude and rate sensors are placed on the LFSS near each AMCD center, and use the desired torque command to generate appropriate magnetic actuator forces. Since the AMCD's are relatively small, this control law approximates the two-level controller discussed in the previous section, wherein additional torque actuators were used for primary attitude control.

Let us consider the first step. The "position" feedback gain G_p in (202) must be redesigned to exclude the feedback of $(\alpha_{ai} - \alpha_s)$. However, the rates $(\dot{\alpha}_{ai} - \dot{\alpha}_s)$ must be zero in steady state, and must be fed back. Thus G_p must be redesigned to control only ε_i so as to keep the rim centers near their nominal positions, and the AMCD rim transverse rotation angles (α_{ai}) are allowed to be nonzero constants in steady state. In the fine-pointing mode, $(\alpha_{ai} - \alpha_s)$ are expected to be very small (on the order of one degree) and it is reasonable to expect that the electromagnetic actuator gap limits will not be exceeded.

The columns of the matrix C_{1i} are given by: $(y_{ij}, -x_{ij})^T$, $j = 1, 2, ..., \ell_i$. Since the actuators of each AMCD are located along a circle, the first two columns of C_{1i} are linearly independent, and columns 3 through ℓ_i can be expressed as linear combinations of the first two columns. That is, C_{1i} can be expressed as:

$$C_{1i} = C_i[I_2 \vdots \Gamma_i] \tag{225}$$

where C_i is the 2×2 matrix consisting of the first two columns of C_{1i} and Γ_i is a $2 \times (\ell_i - 2)$ matirx. Let the position gain be given by:

$$G_p = \text{diag}\left[\begin{pmatrix}\Gamma_1 \\ -I\end{pmatrix}, ..., \begin{pmatrix}\Gamma_\nu \\ -I\end{pmatrix}\right] \tilde{G}_p \text{diag}\left[\begin{pmatrix}\Gamma_1 \\ -I\end{pmatrix}, ..., \begin{pmatrix}\Gamma_\nu \\ -I\end{pmatrix}\right]^T \tag{226}$$

where $\tilde{G}_p = \tilde{G}_p^T > 0$ is an $(\ell - 2\nu) \times (\ell - 2\nu)$ matrix. Using the proportional-plus-derivative control law of Eq. (202), it can be verified by direct substitution that this choice of G_p results in the elimination of the feedback of $(\alpha_{a1}^T - \alpha_s^T, ..., \alpha_{a\nu}^T - \alpha_s^T)^T$ from the control input force f. Letting

$$s = (\varepsilon^T, q^T)^T \tag{227}$$

the resulting closed-loop LFSS/AMCD system can then be written in the form:

$$\begin{bmatrix} \dot{s} \\ \dot{x} \end{bmatrix} = \begin{bmatrix} 0 & 0 \quad I \\ \alpha_1 & \alpha_2 \end{bmatrix} \begin{bmatrix} s \\ x \end{bmatrix} \tag{228}$$

where x is as defined in (191), and α_1, α_2 are appropriately dimensioned matrices. The following theorem addresses the stability of the system in (228).

Theorem 20. *The closed-loop system in eq. (228) is a.s. if $\tilde{G}_p > 0$ and $G_r > 0$.* ■

The proof is similar to that of Theorem 18, and is omitted.

The choice of G_p as in (226) yields a damping enhancement control law which ensures that, as $t \to \infty$, $\varepsilon \to 0$, $q \to 0$ and $\alpha_{ai} - \alpha_s \to$ constant.

For step (ii), (i.e., LFSS attitude control), we can now superimpose force commands on the electromagnetic actuators in order to produce the desired torque for controlling the attitude. The desired attitude control torque is given by (214), wherein the attitude and rate are measured by sensors located on LFSS near AMCD rim centers. Since the AMCD's are small compared to the LFSS, the effect of the control moments generated in this manner will approximate the addition of the primary controller as described previously, and the resulting system will be asymptotically stable. In this configuration, the AMCD's must have sufficiently large momenta in order to exert the magnitudes of torque required to achieve the rigid-body bandwidth specification without exceeding the actuator gap limits.

Robustness of Dissipative Controllers Using AMCD's

We have already established that the stability of dissipative controllers using AMCD's holds regardless of parameter inaccuracies or the number of modes in the model. Furthermore, the electromagnetic force actuators and position sensors are almost perfectly linear (except for actuator saturation). The actuator and sensor bandwidths are very high- on the order of several hundred Hertz; thus actuator time-lag is not expected to be a problem. Nevertheless, it is noteworthy that the previously proven robustness properties of dissipative controllers, namely, tolerance to nonlinearities and first-order actuator dynamics, also hold when AMCD's are used for damping enhancement.

2.7 Remarks on Dissipative Controllers

The statement of stability conditions for linear second-order vector-matrix equations of the type considered in this chapter is generally attributed to Lord Kelvin and Tait [Kel.21]. Their assertion was proved later by Chetayev [Che.61]. The basic concept of dissipative damping enhancement controller was proposed several years ago for controlling elastic aircraft [Wyk.66]. In addition to the references already given in the text, the results presented in this chapter are based on, or are related to the results of: [Bal.79], [Ben.81], [Jos.81], [Jos.83], [Jos.85] and [Jos.86a].

A natural further development in dissipative controllers is the concept of dissipative dynamic compensators. A method for designing dissipative dynamic compensators for damping enhancement has been proposed in [McL.87]. This basically represents a further development in the design of CDEC. In particular, since the elastic mode dynamics (i.e., excluding the rigid-body modes) of LFSS with collocated actuators and sensors have strictly positive real transfer matrix, a sufficient condition for closed-loop stability is that the compensator be positive real. From the Kalman-Yakubovitch lemma [Kal.63], a linear, time-invariant compensator given by

$$\dot{z} = \hat{A}z + \hat{B}y \tag{229}$$

$$u = \hat{C}z \tag{230}$$

has a positive real transfer matrix if and only if there exist symmetric matrices $Q \geq 0$ and $P > 0$ such that

$$\hat{A}^T P + P\hat{A} = -Q \tag{231}$$

and

$$C = \hat{B}^T P \tag{232}$$

A method for designing dissipative dynamic compensators for damping enhancement is given in [McL.87]. Such compensators are robust to modeling errors; i.e., the closed-loop stability is guaranteed in the presence of unmodeled LFSS dynamics or inaccurate knowledge of the parameters. However, development of systematic methods for designing dissipative dynamic compensators for optimal performance, is still an active area of research. Some advances in that direction are reported in [Saf.87], [Loz.88]. Further investigation is also needed for the case when rigid modes are included, and when actuator/sensor nonlinearities and dynamics are present. In summary, design of dissipative dynamic compensators promises to be a fruitful area for further research.

Chapter 3

LQG-Based Controllers

The linear quadratic Gaussian (LQG) controller represents a design technique which has attained considerable maturity since its inception in the fifties and the sixties, and has come to be generally regarded as one of the standard methods. The LQG controller minimizes (in the presence of process and measurement noise) the expected value of a performance function which consists of quadratic functions of the state and control vectors. The infinite duration LQG controller for linear, time-invariant (LTI) systems consists of a constant-gain Kalman-Bucy filter (KBF) followed by a constant linear quadratic regulator (LQR) gain. The resulting compensator is a "model-based compensator" (MBC) because the KBF uses the plant dynamics in its "prediction" part. Thus the plant dynamics must be accurately known in order to design an LQG controller. In the case of large space structures, however, this is almost never the case, because

i) the order of an acceptable dynamic model of the plant is usually very high. It is not practical to implement such high-order KBF.

ii) the parameters (the modal frequencies, damping ratios, and mode-shapes) are not known accurately, especially for higher-frequency modes.

The most obvious way to design an implementable reduced-order LQG controller is to truncate the model after a certain number of elastic modes, and to base the design on the reduced order "design" model. However, mere truncation is often not the best way of obtaining a reduced-order model. Balanced realization [Moo.81, Per.82] represents a better method for order reduction. The basic idea is to find a similarity transformation such that the controllability and observability grammians of the transformed plant dynamics are equal and diagonal. The transformation is chosen to arrange the elements (called the "Hankel singular values") of the controllability/observability grammian in descending order, and only the first k states, which correspond to the k largest Hankel singular values, are retained. A balanced realization can be obtained for any asymptotically stable system. Therefore, it can be used for the flexible-body dynamics of LFSS. Another related method for obtaining reduced-order models is the optimal Hankel norm approximation [Glo.84]. This method gives a smaller bound on the error of approximation, but has an interesting feature, namely, the reduced-order model obtained is not generally strictly proper even though the plant is strictly proper. The stable factorization approach [Vid.85] wherein the plant transfer matrix is expressed as a fraction of stable and proper numerator and denominator transfer matrices, offers yet another method for order reduction ([Liu.87], [Vid.87]). One problem with balanced realization, Hankel norm, and stable factorization approximations is that the transformations used couple all the modes and the state variables of the reduced model do not correspond to the individual elastic modes. An alternate method is to assess the relative controllability/observability of each elastic mode by applying the Popov-Belevich-Hautus (PBH) rank test [Kai.80] to the balanced realization. In particular, we examine the reciprocal condition numbers (defined as ratio of smallest to largest singular value) of the matrix $[sI - A : B]$ at $s = \lambda_i(A)$. We then rank the elastic modes according to their reciprocal condition numbers rather than the Hankel singular values, and retain the first k modes with the largest reciprocal condition numbers [Jos.86b]. The underlying assumption is that the ranking of the modes according to controllability and observability is the same. (This was found to be the case in many numerical examples involving LFSS). Another order reduction method is the "q-covariance" approximation [You.85], in which the first q Markov parameters, as well as the output covariances of the full-order and reduced-order models are equal. In this book, we do not investigate methods for order reduction, but rather study the effect of using reduced order models for controller design.

No matter which method is used to obtain the reduced-order model, the closed-loop stability can no longer be guaranteed because of the modeling errors, which include the control and observation "spillovers" discussed in Chapter 1. In addition, inaccurate knowledge of the parameters also contributes to the modeling errors. Furthermore, we must ensure stability in the presence of actuator and sensor nonlinearities, and failures. In this chapter, we investigate the design of reduced-order LQG-type controllers which are robust to modeling errors and actator/sensor nonlinearities. As stated previously, we define a controller to be robust if it can maintain at least stability, and if possible, the performance, in the presence of these problems. The organization of this chapter is as follows: Sections 3.1 and 3.2 deal with robustness to modeling errors. Section 3.1 investigates time-domain LQG design approaches for LFSS. The problems in designing standard LQG-type controllers for LFSS, and some methods of overcoming them are discussed. Section 3.2 discusses the application of multivariable frequency-domain methods for designing MBC's for LFSS. The stability in the presence of actuator and sensor nonlinearities is investigated in Sec.3.3.

3.1 Time-Domain LQG Design Approaches

In this section we outline some approaches for reduced-order LQG- type controller design for LFSS. For convenience we shall use the rotational motion model described in eqs. (27)-(33) in Chapter 2, wherein we allow only torque inputs. We assume that the sensors are generally not at the same locations as the actuators, so that the attitude and rate sensor outputs are given by:

$$y_p = \Gamma_s p + w_p \tag{1}$$

$$y_r = \Gamma_s \dot{p} + w_r \tag{2}$$

where

$$\Gamma_s^T = \begin{bmatrix} I_3 & I_3 & \dots & I_3 \\ \Phi_1^T & \Phi_2^T & \dots & \Phi_\ell^T \end{bmatrix}_{(n_q+3)\times 3\ell} \tag{3}$$

Φ_i's being the mode-slope matrices at the ℓ sensor locations (Each sensor is assumed to be 3-axis). w_p and w_r are the attitude and rate measurement noise vectors, assumed to be zero-mean and white.

The equations of motion can be expressed in the state-space form as:

$$\dot{x} = Ax + Bu + v \tag{4}$$

where

$$x = \left(\alpha^T, \dot{\alpha}^T, q_1, \dot{q}_1, q_2, \dot{q}_2, \ldots, q_{n_q}, \dot{q}_{n_q}\right)^T \tag{5}$$

A and B are defined appropriately, and v is the process noise, assumed to be zero-mean and white. The corresponding sensor equation is:

$$y = Cx + w \tag{6}$$

where $y = [y_p^T, y_r^T]^T$. Thus the designer's best knowledge of the plant is described by eqs. (4)-(6), which is called the "evaluation model" [Ske.78].

As stated previously, the order of the evaluation model is usually high; therefore, the full order LQG controller based on that model would be impractical. We assume that a reduced-order model has been obtained in such a manner that it has the same structure as the plant; i.e., it consists of the rigid-body modes and some of the elastic modes of the evaluation model. Thus the reduced-order ("design") model, is given by:

$$\dot{x}_d = A_d x_d + B_d u + v_d \tag{7}$$

$$y_d = C_d x_d + w_d \tag{8}$$

The uncontrolled or residual mode dynamics are then given by:

$$\dot{x}_r = A_r x_r + B_r u + v_r \tag{9}$$

$$y_r = C_r x_r + w_r \tag{10}$$

The LQG performance function based on the design model is:

$$J = \lim_{t_f \to \infty} \frac{1}{t_f} \mathcal{E} \int_0^{t_f} [x_d^T(t) Q x_d(t) + u^T(t) R u(t)] dt \tag{11}$$

where $Q = Q^T \geq 0, R = R^T > 0$, are the state and control weighting matrices. The LQG controller which minimizes J in (11) is given by:

$$u = G\hat{x}_d \tag{12}$$

$$G = -R^{-1} B_d^T P \tag{13}$$

where

$$A_d^T P + P A_d - P B_d R^{-1} B_d^T P + Q = 0 \tag{14}$$

$$\dot{\hat{x}}_d = A_d \hat{x}_d + B_d u + H(y_d - C_d \hat{x}_d) \tag{15}$$

$$H = \Sigma C_d W_d^{-1} \tag{16}$$

$$A_d \Sigma + \Sigma A_d^T - \Sigma C_d^T W_d^{-1} C_d \Sigma = -V_d \tag{17}$$

where V_d, W_d are the covariance intensities of the process and measurement noise, G and H denote the LQR and KBF gain matrices, and P, Σ denote the corresponding Riccati matrices. In actual implementation, however, it is not possible to measure y_d for use in (15); the actual sensor output is:

$$y = C_d x_d + C_r x_r + w \tag{18}$$

and the KBF equation is:

$$\dot{\hat{x}}_d = A_d\hat{x}_d + B_d u + H(y - C_d\hat{x}_d) \tag{19}$$

The term "$B_r u$" in eq. (9) represents the control spillover, and the term "$C_r x_r$" in eq. (18) represents the observation spillover. Assuming that the parameters are accurately known, the closed-loop stability would be guaranteed if either or both of the spillover terms were zero. (Obviously, zero control spillover implies no excitation of the residual modes by the control input, and zero observation spillover implies no feedback of x_r into x_d -dynamics). Because of continuity, the stability should also hold for sufficiently small values of the spillover "gain" terms, quantized by the spectral norms: $\| B_r G \|_s$ and $\| H C_r \|_s$. One way of reducing the spillover gains is to reduce the magnitudes of the regulator gain G and the Kalman gain H, which can be accomplished by reducing the corresponding state vector weighting matrices in the LQR and KBF designs. Of course, the performance will also deteriorate because of reduction in these gains.

Another way of reducing the observation spillover is to use an estimator which is of order higher than that of the regulator. If a full order KBF is used, the observation spillover will be zero. However, this tactic does not gain us much in terms of implementation because the order of the controller (which is the same as the order of the KBF) would still be high.

One of the standard methods used in conventional (relatively rigid) spacecraft controller design was to insert low-pass filters in the paths of the sensor outputs. In general this method is not effective for LFSS [Jos.79] because the flexibility is quite dominant; i.e., the elastic mode shapes (and slopes) are much larger in magnitude than those for conventional spacecraft.

An approach to reducing the observation spillover consists of modeling it as a linear combination of some (basis) functions of time t, and estimating it without utilizing the knowledge of residual mode dynamics. Controllers using such estimators, while not requiring the knowledge of the residual dynamics, do increase the controller order. Skelton and Likins proposed such an approach wherein the spillover was modeled using

Chebychev polynomials [Ske.78]. Joshi and Groom [Jos.79] proposed the use of polynomials in t for the same purpose, but found that it was not effective in improving the closed-loop stability.

Sesak et al [Ses.79] proposed the "model error sensitivity suppression (MESS)" method in which the knowledge of the residual dynamics is used to modify the LQR weighting matrices so as to minimize the effect of spillovers. Ignoring noise, the steady-state version of the residual dynamics is given by:

$$\dot{\bar{x}}_r = 0 = A_r \bar{x}_r + B_r u \tag{20}$$

which implies

$$\bar{x}_r = -A_r^{-1} B_r u \tag{21}$$

The idea is to include a quadratic penalty: $\bar{x}_r^T Q_r \bar{x}_r$ in the LQG performance function, so that the modified performance functon becomes [using (21)]:

$$J = \lim_{t_f \to \infty} \frac{1}{t_f} \mathcal{E} \int_0^{t_f} \left[x_d^T Q x_d + u^T (R + B_r^T A_r^{-T} Q_r A_r^{-1} B_r) u \right] dt \tag{22}$$

This approach is equivalent to designing the "slow" regulator using the singular perturbation approach wherein the residual dynamics are considered to be the fast dynamics. An alternative method is to include a quadratic penalty on $(B_r u)$, which is the actual excitation going into the residual mode dynamics. The ultimate result of these methods is modification of the input weighting matrix R. This approach was found to give significantly better results than mere truncation of the model [Ses.79; Jos.79].

The methods described in this section assume the knowledge of the design model. With this assumption, we next consider the problem of quantizing the tolerable spillover terms so that the closed-loop stability is assured.

3.1.1 Stability Bounds on Spillover

We consider a general observer (rather than a KBF), described by:

$$\dot{z} = A_z z + B_z u + Ky \tag{23}$$

where z is the estimated value of x_d, and A_z, B_z, K are constant matrices. The input is given by:

$$u = Gz \tag{24}$$

The resulting closed-loop system (ignoring noise terms) can be expressed as:

$$\begin{bmatrix} \dot{x}_1 \\ \dot{x}_r \end{bmatrix} = \begin{bmatrix} A_1 & 0 \\ 0 & A_r \end{bmatrix} \begin{bmatrix} x_1 \\ x_r \end{bmatrix} + \begin{bmatrix} 0 & \delta \\ \beta & 0 \end{bmatrix} \begin{bmatrix} x_1 \\ x_r \end{bmatrix} \tag{25}$$

where $x_1 = \left(x_d^T, z^T\right)^T$,

$$A_1 = \begin{bmatrix} A_d & B_d G \\ KC_d & A_z + B_z G \end{bmatrix}, \delta = \begin{bmatrix} 0 \\ KC_r \end{bmatrix}, \beta = [0 \, B_r G] \tag{26}$$

Denote $x = (x_1^T, x_r^T)^T$ and $A = \text{diag}\ (A_1, A_r)$. Note that A_1 and A_r are strictly Hurwitz matrices because the controlled part (design model and observer with stabilizing feedback gain) is stable, and the residual modes have some inherent damping. Given appropriately dimensioned symmetric positive definite matrices Q, Q_1, and Q_r, there exist symmetric positive definite matrices P, P_1, and P_r such that

$$A^T P + PA = -Q \tag{27}$$

$$A_1^T P_1 + P_1 A_1 = -Q_1 \tag{28}$$

$$A_r^T P_r + P_r A_r = -Q_r \tag{29}$$

If Q_r is block-diagonal consisting of 2×2 blocks (one block for each mode), it is interesting to note that eq. (29) for the residual system decomposes into a set of *uncoupled* 2×2 Lyapunov equations which can be readily solved in *closed-form.*

The following theorem gives a sufficient condition for asymptotic stability of the system in (25).

Theorem 1. *The system (25) is asymptotically stable (a.s.) if*

$$\max(\| KC_r \|_s, \| B_r G \|_s) < \lambda_m(Q)/[2\lambda_M(P)] \triangleq \beta_1 \tag{30}$$

Proof. Consider the Lyapunov function

$$V(x) = x^T P x \tag{31}$$

Then

$$\dot{V}(x) = -x^T Q x + 2x^T P \Delta \dot{x} \tag{32}$$

where

$$\Delta = \begin{bmatrix} 0 & \delta \\ \beta & 0 \end{bmatrix} \tag{33}$$

Since

$$2x^T P \Delta x \leq 2 \| x \|^2 \lambda_M(P) \| \Delta \|_s \tag{34}$$

$\dot{V}$ is negative definite if

$$\| \Delta \|_s < \lambda_m(Q)/[2\lambda_M(P)] \tag{35}$$

From (33), it can be proved that

$$\| \Delta \|_s = \max(\| B_r G \|_s, \| KC_r \|_s) \tag{36}$$

Inequality (30) can be obtained by using (36) in (35). ■

We now obtain another sufficient condition for stability using a weighted sum of scalar Lyapunov functions for composite systems. We first prove the following result.

Lemma. *The system (25) is a.s. if there exist positive scalars δ_1 and δ_2 such that*

$$\delta_1 \lambda_M(P_1) \parallel KC_r \parallel_s + \delta_2 \lambda_M(P_r) \parallel B_r G \parallel_s < \sqrt{\delta_1 \delta_2 \lambda_m(Q_1) \lambda_m(Q_r)} \tag{37}$$

Proof. Eq. (25) consists of two interconnected subsystems. Consider the Lyapunov functions:

$$V_1(x_1) = x_1^T P_1 x_1 \quad \text{and} \quad V_2(x_r) = x_r^T P_r x_r \tag{38}$$

where P_1 and P_r are described previously and satisfy (28) and (29). Then we have the following:

$$\lambda_m(P_1) \parallel x_1 \parallel^2 \leq V_1(x_1) \leq \lambda_M(P_1) \parallel x_1 \parallel^2 \tag{39}$$

$$\dot{V}_1 = -x_1^T Q_1 x_1 \leq -\lambda_m(Q_1) \parallel x_1 \parallel^2 \tag{40}$$

$$\left| \frac{\partial V_1}{\partial x_1} \right| \leq \parallel 2P_1 x_1 \parallel \leq 2\lambda_M(P_1) \parallel x_1 \parallel \tag{41}$$

Similar inequalities can be obtained for the second subsystem. Using Theorem 2 of [Mic.72], the sufficient condition (37) can be obtained. ■

We next prove the following reult.

Theorem 2. *System (25) is a.s. if*

$$\parallel KC_r \parallel_s \parallel B_r G \parallel_s < \frac{\lambda_m(Q_1)\lambda_m(Q_r)}{4\lambda_M(P_1)\lambda_M(P_r)} \triangleq \beta_2^2 \tag{42}$$

Proof. Letting $r = \sqrt{\delta_2/\delta_1}$, (37) can be expressed as:

$$r^2\lambda_M(P_r) \parallel B_rG \parallel_s - r\sqrt{\lambda_m(Q_1)\lambda_m(Q_r)} + \lambda_M(P_1) \parallel KC_r \parallel_s < 0 \tag{43}$$

To get the least conservative sufficient condition, we minimize the left hand side of (43) with respect to r. This yields (42). ∎

The sufficient condition (42) was obtained using a weighted sum of scalar Lyapunov functions for interconnected systems, which has been shown in [Mic.72] to be generally less conservative than the vector Lyapunov function approach in [Bai.66]. Also note that (42) can be obtained independently using the results in [Gru.73] for interconnected systems.

The sufficient conditions for stability derived above are generally quite conservative, which is an attribute common to Lyapunov-based methods. The choice of matrices Q, Q_1 and Q_r is arbitrary. In numerical evaluations, we found that the condition (42) of Theorem 2 was far less conservative than condition (30) of Theorem 1. If Q in eq. (27) is block-diagonal with two blocks corresponding to the 1-system and the r-system, eq. (27) gives two uncoupled Lyapunov equations, namely, (28) and (29). In that case, it can be verified after some straightforward reasoning that

Numerator of $\beta_2 \geq$ Numerator of β_1 , and

Denominator of $\beta_2 \leq$ Denominator of β_1 .

Therefore, $\beta_2 \geq \beta_1$, and condition (42) is less conservative than condition (30), at least for this particular type of Q matrix.

3.1.2 Controller Design for Hoop/Column Antenna

Let us consider the attitude controller design problem for the 122 meter hoop/ column antenna which was described in Chapter 1. We investigated the design of a dissipative controller for this antenna in Chapter 2. Herein we present an LQG controller design based on a reduced-order model obtained by simple truncation. The design objectives are to obtain sufficiently high closed-loop bandwidth, sufficiently rapid vibration damping, and root mean square (RMS) pointing errors smaller than the specifications. As in the case of dissipative controller design, four 3-axis torque actuators and attitude and rate sensors at the same locations (Fig. 5 of Chapter 2) are assusmed.

The first design itereation consisted of using the simplest "design" model (i.e., only three rigid modes), performing the best possible design, and computing the RMS errors (as discussed in Chapter 2). The LQR weighting matrices were determined by trial and error so as to give the desired closed-loop rigid-body frequency (ω_α) and damping ratio (ρ_α). The same nominal desired values of ω_α and ρ_α were selected (as in Chapter 2); i.e., ω_α = 0.02, 0.1, and 0.25 rad/sec, and ρ_α = 0.7. While designing the KBF, it was found that the use of the actual values of the noise covariance intensities gives an extremely slow KBF; therefore, it was decided to use the KBF weighting matrices only as design parameters (with no statistical meaning), and they were adjusted to give a filter with closed-loop frequencies corresponding to the rigid modes which were 3-4 times faster than those for the regulator, and with damping ratio of approximately 0.7. It became immediately apparent that the 6th order LQG controller (based on the rigid design model) cannot achieve ω_α beyond about 0.01 rad/sec without causing spillover-induced instability. Therefore the first mode was added to the design model, and the same procedure was repeated. This design model failed to increase the bandwidth beyond 0.015. In fact, at least the first three modes had to be included in the design model in order to obtain ω_α greater than 0.1 rad/sec. The nominal design model was chosen to include the first six elastic modes. For all the cases where the design model included one or more elastic modes, the LQR weighting matrices (Q and R) were adjusted to obtain a damping ratio of approximately 0.5 for the corresponding closed-loop eigenvalues. (The damping ratio for an elastic mode can be increased by increasing the weight on the corresponding modal velocity $\dot{q}_i$). The corresponding KBF closed-loop eigenvalues were designed to be 3-4 times faster by appropriately increasing the weights corresponding to $\dot{q}_i$ in the filter design. For each ω_α, the LQR weights corresponding to the modal velocities were increased at each step by a factor of 10, and the nominal performance was computed similar to the dissipative controller case. The stability of the full-order plant was checked at each step. For this 18th order (rigid plus six elastic mode) compensator design, the spillovers did not cause appreciable destabilizing effects; the

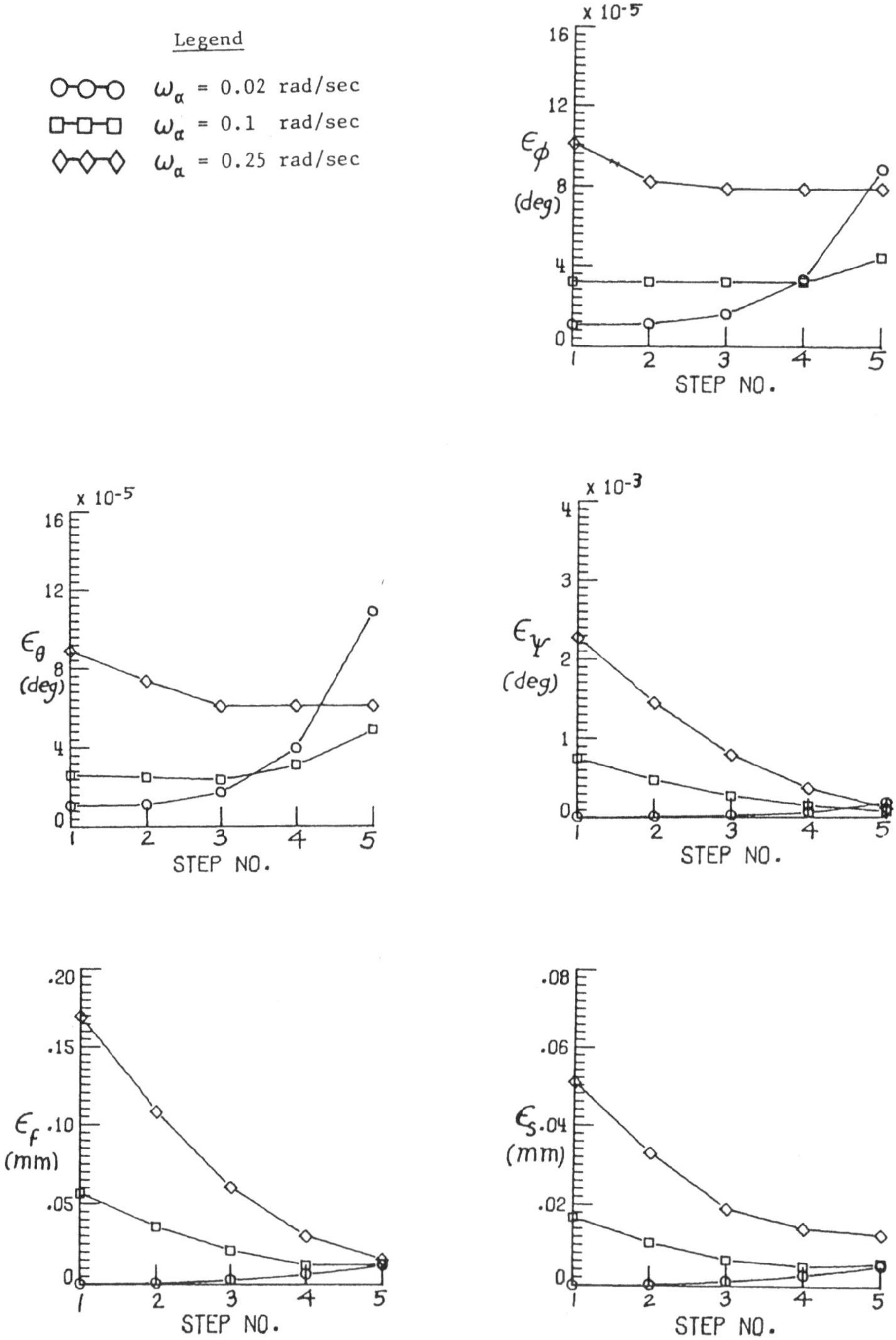

Figure 1. Performance of LQG controller

Legend

○─○─○ $\omega_\alpha = 0.02$ rad/sec

□─□─□ $\omega_\alpha = 0.1$ rad/sec

◇─◇─◇ $\omega_\alpha = 0.25$ rad/sec

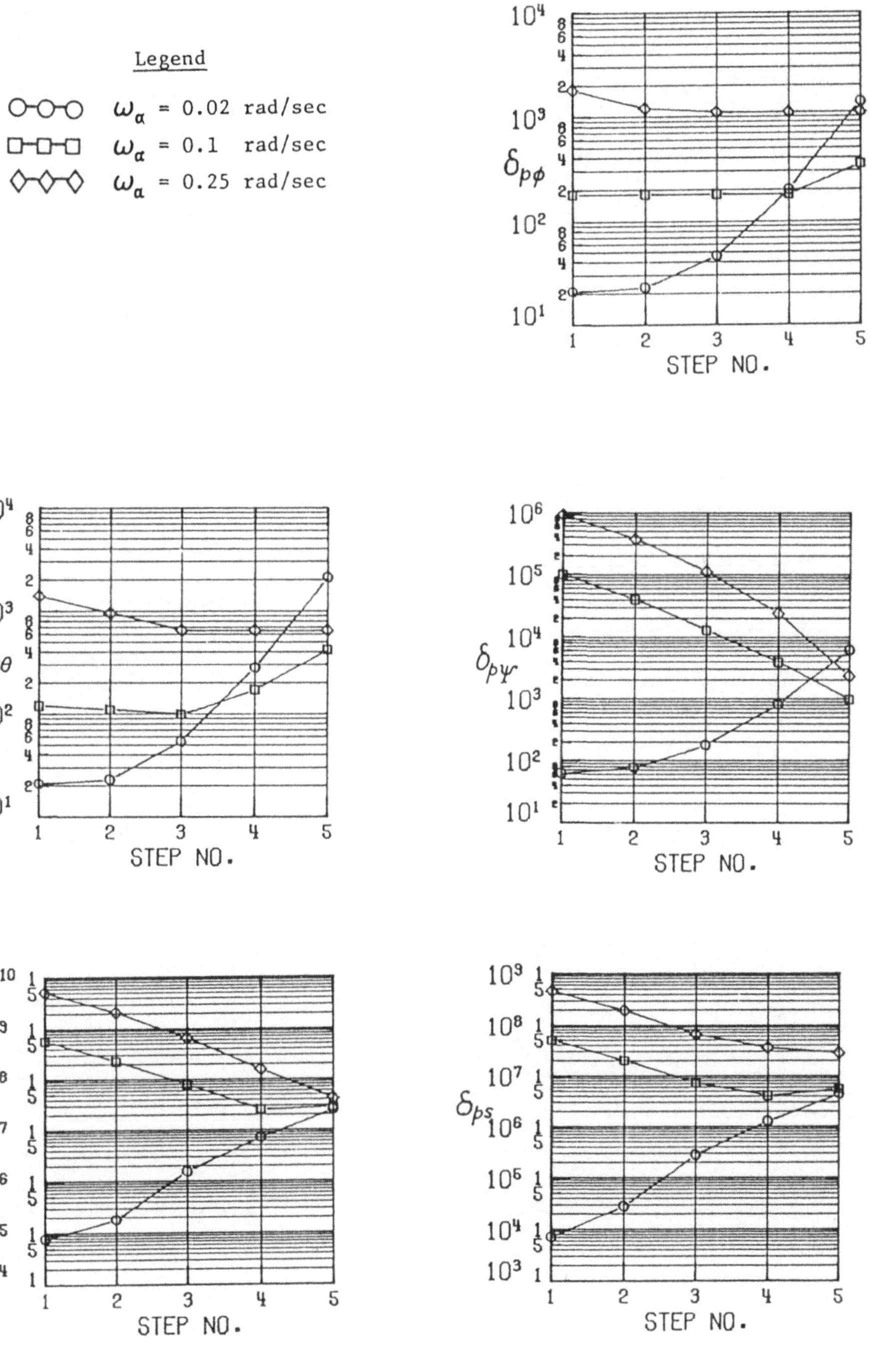

Figure 2. Performance coefficient δ_P

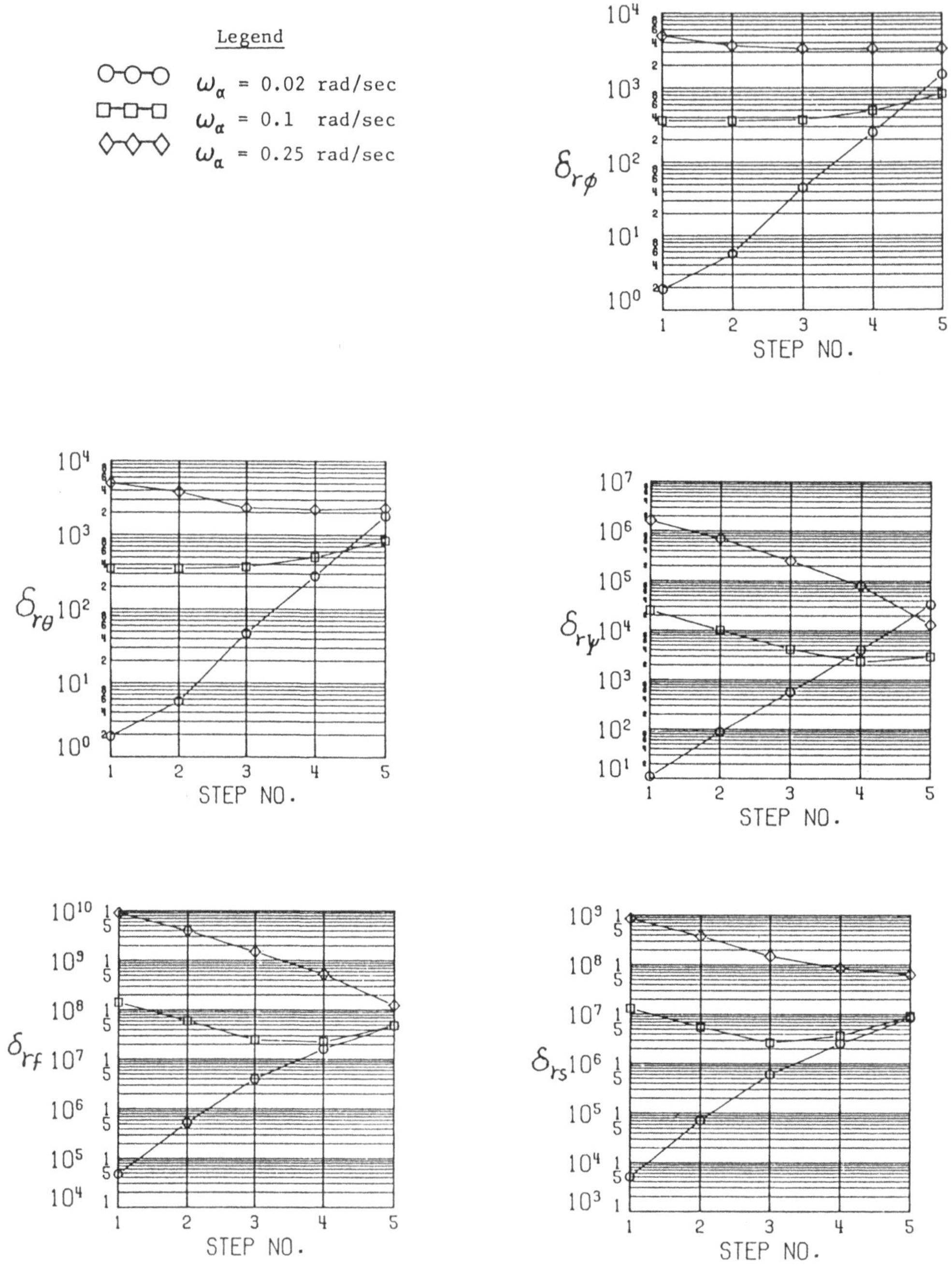

Figure 3. Performance coefficient δ_r

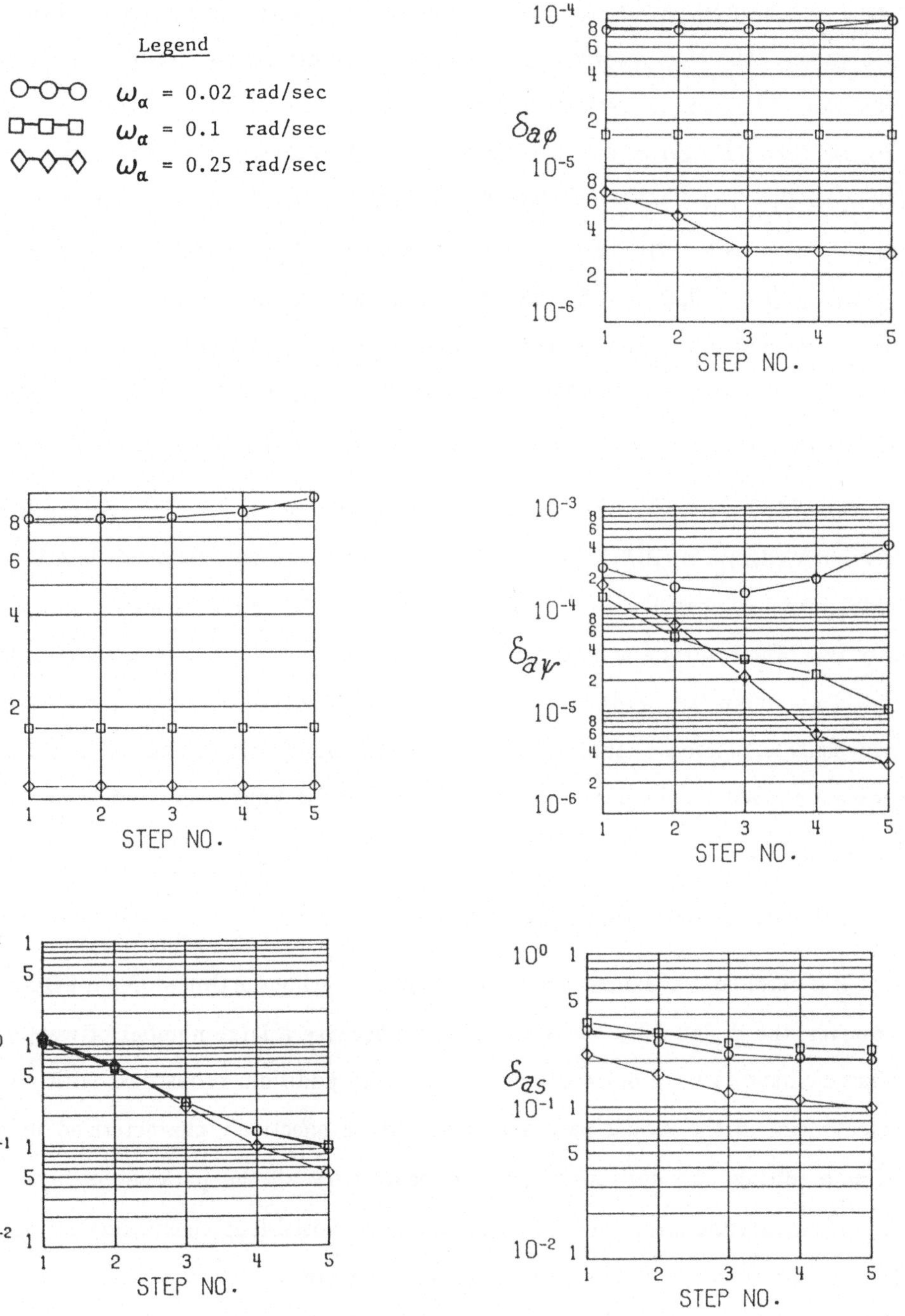

Figure 4. Performance coefficient δ_a

closed-loop damping ratios for the residual modes changed very little from their open-loop values. Therefore, it was not necessary to use any special techniques for spillover reduction. The nominal performance is plotted in Figure 1 for $\omega_\alpha = 0.02$, 0.1, and 0.25 rad/sec. The RMS errors are significantly lower than those for the dissipative controller (Chapter 2). The RMS error data for the LQG controller were parametrized in the same fashion as for the dissipative controller case. Figures 2-4 show the coefficients $\delta_{pi}, \delta_{ri}, \delta_{ai}, (i = \phi, \theta, \psi, f, s)$ for the LQG controller. A comparison with the δ's for the dissipative controller indicates that δ_{pi} and δ_{ri} are much lower for the LQG controller, while δ_{ai} appears to be roughly the same. The LQG feedback gains were also much smaller than the dissipative controller feedback gains (by a factor of 100).

The LQG controller was found to be relatively insensitive to inaccuracy in parameter knowledge; a $\pm$ 10% change in ω_i and ρ_i caused only about 4% deterioration in RMS errors. It was also found that bandwidth of 0.25 rad/sec was achievable using only two, 3-axis torque actuators. With only one actuator, the best achievable bandwidth was about 0.12 rad/sec. However, the sensitivity to errors in the design model (especially ω_i) increased subtantially. It was also found that adding more than the first six elastic modes to the design model did not result in appreciable improvement in the performance.

Remarks on Time-Domain LQG Design

It is possible to obtain an acceptable controller using the time-domain LQG mehtod. However, the design relies on trial and error because a large number of weighting parameters have to be simultaneously adjusted. In addition, robustness with respect to inaccuracies in the modal parameters cannot be effectively characterized in a time-domain setting; the only way seems to be to vary all the parameters in a random manner within the given limits, and to check the closed-loop eigenvalues for each case. Multivariable frequency-domain methods offer a more systematic approach to overcome these difficulties, and are discussed in the next section.

3.2 Multivariable Frequency-Domain Methods

Classical frequency-domain methods have been the main controller design tools for single-input single-output (SISO) systems for several decades. These methods utilize techniques such as the root locus, Bode plots, Nyquist plots, and Nichol's chart. Classical design specifications include: the shape of the closed loop frequency response; the bandwidth; the gain margin, and the phase margin. The first two items are indicative of the performance, while the latter two are indicators of robustness.

Numerous attempts have been made to extend the classical scalar design techniques to the multi-input multi-output (MIMO) case. Two of the best-known of such extensions are the inverse Nyquist array method [Ros.74] and the characteristic loci method [Mac.77]. These mehods are based on converting the MIMO design problem into a number of SISO problems. This is accomplished by using pre and post compensators to make the system almost diagonal. However, these mehtods have been shown to lack robustness; i.e., an apparently good sequence of scalar designs may yield a multivariable system with poor robustness [Doy.81].

An alternative approach is to use singular values of the transfer function matrix to "scalarize" the MIMO system. A singular value of a complex matrix $P(\omega)$ is defined as:

$$\sigma[P(\omega)] = \{\lambda[P^*(\omega)P(\omega)]\}^{1/2}$$

where $P^*(\omega)$ denotes the conjugate transpose of $P(\omega)$. The largest singular value, denoted by $\bar{\sigma}[P]$, is a measure of the magnitude or "gain" of P, and the smallest singular value, denoted by $\underline{\sigma}[P]$, is a measure of the distance between P and the singular matrix "nearest" to P. The advantage of using singular value methods is that they can handle both performance and robustness issues. (See [Rid. 86] for a nice tutorial on multivariable systems using singular value methods).

In this section we investigate the application of a class of multivariable frequency-domain methods for the design of robust compensators for LFSS using singular values of the transfer matrices. One of the methods proposed in the literature using MIMO frequency domain techniques is the LQG/loop transfer recovery (LTR) method [Ath.86; Doy.81; Ste.87]. We consider the LQG/ LTR method, modify it for the LFSS problem, and apply it for synthesizing an attitude control system for the 122 m diameter hoop/column antenna. From the results abtained, this method is found to give a controller that is robust to unmodeled elastic dynamics, but not to uncertainty in the design model.

3.2.1 The LQG/LTR Method

The full state feedback LQ regulator has been shown to have desirable robustness properties, namely, 60° phase margin, infinite gain margin, and tolerance to actuator nonlinearities in the $(1/2, \infty)$ sector [Saf.77]. In the frequency domain, these gain and phase margin characteristics imply desirable frequency response (or "loop shape"). However, the LQR is not implementable in most practical problems because it requires full state feedback. The problem then, is to design an implementable controller which has the desired performance and robustness properties. An LQG controller is certainly implementable, but has no guaranteed robustness properties [Doy.78]. It is intuitively apparant to see that if the KBF (observer) is made faster and faster, the LQG controller would approach the LQR in the limit; i.e., it would "recover" the desirable LQR loop-shape. The LQG/LTR method is based on this asymptotic property, [Doy.81]. The mathematical dual of this asymptotic property is that the KBF loop-shape can be "recovered" by increasing the LQR gain [Kwa.72a]. This asymptotic recovery property holds only for minimum phase systems; i.e., only for those systems having transmission zeros in the open left-half plane. (One interpretation of the LQG/LTR method is that it approximately inverts the system, and replaces it with the desired loop transfer function. i.e., $C = P^{-1}P_{\text{desired}}$, and unstable pole-zero cancellations are not allowed). The LQG/LTR method also uses frequency domain inequalities involving matrix norm bounds (i.e., singular values) of certain combinations of transfer matrices of the nominal plant, the controller, and the plant uncertainty [San.78; Cru.81; Vid.85]. The resulting controller is a "model-based compensator (MBC)" because the KBF uses the nominal model of the plant. The LQG/LTR method has been applied to diverse systems such as power systems [Cha.84] and aircraft engine control [Kap.83]. The design philosophy is to use a low-frequency "design model" of the plant and a high- frequency characterization of the modeling errors. The modeling errors for LFSS result from i) unmodeled dynamics, and ii) lack of accurate knowledge of the parameters (i.e., frequencies, mode shapes, and damping ratios). Typically, the parameter errors for finite element models increase substantially with increasing modal frequency.

The application of the LQG/LTR method to LFSS involves synthesizinig a sequence of low order compensators, obtained by treating a sequence of design models which include increasingly higher frequency modes. The final design model chosen is the first one in this sequence which allows the performance/robustness objectives to be met. In this sequence of design models, the first one consists of the rigid body modes only. Subsequent design models are obtained by the succesive addition of flexible modes.

Frequency-Domain Mathematical Model

We use the frequency-domain version of the mathematical model described by eqs. (27)-(35) of Chapter 2. We assume that the controlled quantity is the 3-axis attitude at one particular sensor location, and that the actuation will be done by m_T, 3-axis torque actuators. The 3-axis attitude y_p at a sensor (e.g. a star tracker) location is given by:

$$y_p = \alpha + \Psi q \tag{44}$$

where Ψ is the $3 \times n_q$ mode-slope matrix at the sensor location. If an attitude rate sensor (e.g. a rate gyro) is used, the sensor output y_r is given by an equation similar to (44), except that α and q are replaced by $\dot{\alpha}$ and $\dot{q}$, respectively. The transfer function matrix from the torque inputs to the 3×1 attitude output at a single sensor location is given by:

$$P'(s) = P_1(s) + P_2(s) \tag{45}$$

where

$$P_1(s) = \left(J_s/s^2\right)[I : I : ..I] \tag{46}$$

$$P_2(s) = \sum_{i=1}^{n_q} (\Psi_i \Phi_i^T)/(s^2 + 2\rho_i \omega_i s + \omega_i^2) \tag{47}$$

J_s is the 3×3 moment of inertia matrix, and Ψ_i and Φ_i are the mode-slope matrices for the ith mode at the sensor and actuator locations respectively.

The basic design objectives for the control systems are: (1) to obtain sufficiently high bandwidth (which roughly corresponds to closed-loop frequencies of the rigid body modes) and satisfactory closed loop damping ratios for rigid-body and strucutural modes; (2) to obtain robustness to unmodeled dynamics. The bandwidth is defined in the frequency domain setting as the frequency ω_b at which the closed-loop transfer function magnitude drops below unity; i.e., $\underline{\sigma}[G_{c\ell}(j\omega)] \leq 1$, where $G_{c_\ell}(s) = P(s)C(s)[I + P(s)C(s)]^{-1}$ (Figure 5). In the higher frequency region (i.e., near and beyond ω_b), the loop gain PC rolls off, i.e., $\bar{\sigma}[PC] << 1$. Therefore, $\sigma[G_{c_\ell}] \approx \sigma[PC]$, and it suffices to examine the loop gain rather than the closed-loop transfer function.

The Design Procedure

The LQG/LTR method basically consists of the following steps:

(1) Define a "design" model of the nominal plant which is an acceptable low frequency representation. Define the high frequency uncertainty (robustness) barrier, i.e., the upper limit on the uncertainty in the high frequency range. Define the low frequency performance barrier, which is the envelope of the smallest acceptable loop gain in the low frequency region.

(2) Define the "target feedback loop-shape", which is done by designing a full state feedback compensator based on the steady state Kalman-Bucy filter (KBF). This assumes that the loop is broken at the plant output, which is consistent with the fact that we need good output tracking. Adjust the weighting matrices in the KBF design until its frequency response meets the robustness specifications at high frequencies and bandwidth specification at low frequencies.

(3) Design an LQ regulator to asymptotically "recover" the frequency response obtained in step 2.

(4) Verify stability, and robustness, performance for the entire closed-loop system.

The design model consists of the rigid-body modes plus the first few elastic modes. (The design model must always include the three rigid body modes). The uncertainty

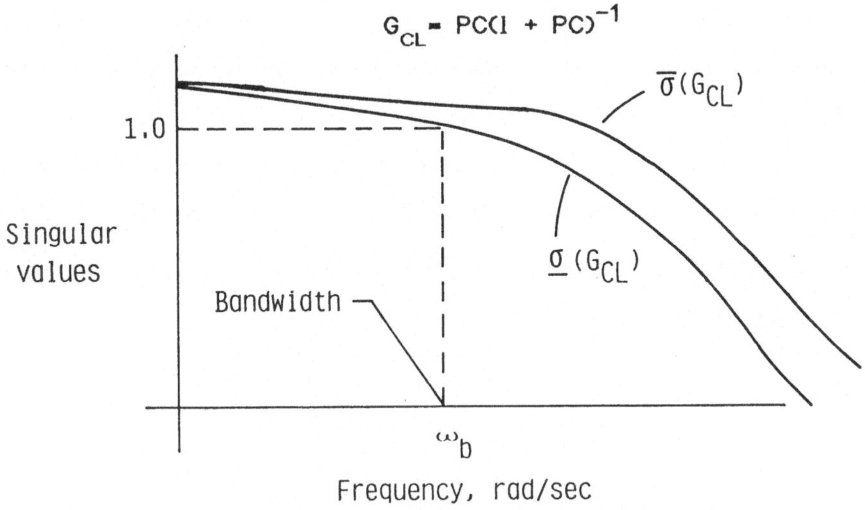

Figure 5. Definition of bandwidth

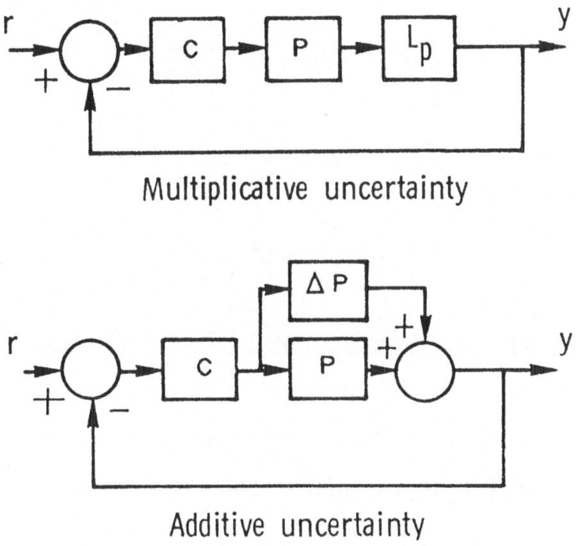

Figure 6. Definition of uncertainty

barrier is a measure of the plant uncertainty at high frequencies. The plant uncertainty can be represented as either multiplcative or additive uncertainty (Fig. 6). The LQG/LTR approach requires the characterization of the uncertainty in terms of a frequency- dependent upper bound. Frequency domain sufficient conditions are used to test the robustness in the presence of uncertainties within that bound.

For the case of multiplicative uncertainty $L_p(s)$ of Figure 6a, with the loop broken at the plant output, the plant transfer matrix is given by:

$$P'(s) = L_p(s)P(s) \tag{48}$$

The closed-loop system is stable if [Vid.85]

$$\bar{\sigma}[L_p - I] < 1/\bar{\sigma}[G_{c\ell}] = 1/\bar{\sigma}[PC(I + PC)^{-1}] \tag{49}$$

wherein the argument $j\omega$ has been dropped for convenience. Using the fact that

$$G_{c\ell}^{-1} = [PC(I + PC)^{-1}]^{-1} = (I + PC)(PC)^{-1} = I + (PC)^{-1} \tag{50}$$

and that $\bar{\sigma}[G] = \underline{\sigma}^{-1}[G^{-1}]$, (49) is equivalent to:

$$\bar{\sigma}[L_p - I] < \underline{\sigma}[I + (PC)^{-1}] \tag{51}$$

where $P(s)$ and $C(s)$ are the design model (plant) and compensator transfer matrices, and $\bar{\sigma}[.]$ and $\underline{\sigma}[.]$ denote the largest and the smallest singular values of the argument matrix respectively. At high frequencies, assuming $\| L_p(j\omega) \| >> 1$ and $\| P(j\omega)C(j\omega) \| << 1$, (49) approximately yields

$$\bar{\sigma}(PC) < 1/\bar{\sigma}[L_p] \tag{52}$$

The "uncertainty (or robustness) barrier" is an upper bound $\ell_m(\omega)$ on $\bar{\sigma}(L_p)$. The system is stable in the presence of such unstructured uncertainties if $\bar{\sigma}[PC] < l_m^{-1}(\omega)$ at high frequencies.

When the additive uncertainty formulation (Figure 6b) is used, the plant is expressed as:

$$P'(s) = P(s) + \Delta P(s) \tag{53}$$

where P and ΔP represent the nominal plant (design model) and the additive perturbation, respectively. A sufficient condition for stability robustness is given by [Vid.85]:

$$\bar{\sigma}[\Delta P] < 1/\bar{\sigma}[C(I + PC)^{-1}] \tag{54}$$

Using Schwarz inequality and standard properties of singular values, another (more conservative) sufficient condition can be obtained:

$$\bar{\sigma}[\Delta P] < \frac{\underline{\sigma}[I + PC]}{\bar{\sigma}[C]} \tag{55}$$

(Condition (55) appears to have been commonly used in much of the literature, although condition (54) is less conservative). At high frequencies, assuming $\bar{\sigma}(PC) << 1$, (55) approximately yields

$$\bar{\sigma}[C] < 1/\bar{\sigma}[\Delta P] \tag{56}$$

That is, the compensator must roll off sufficiently rapidly at high frequencies. The main feature of the LQG/LTR approach is that a target feedback loop, i.e., a full state compensator based on KBF, is first designed, and then a compensator $C(s)$ is obtained such that the loop gain PC approximates the target loop frequency response. Therefore, any loop shaping should involve the product PC rather than C alone. If P is a square matrix,

$$C = P^{-1}(PC)$$

$$\bar{\sigma}(C) \leq \bar{\sigma}(P^{-1})\bar{\sigma}(PC) \tag{57}$$

or

$$\bar{\sigma}(C) \leq \underline{\sigma}^{-1}(P)\bar{\sigma}(PC) \tag{58}$$

Using (58) and (55), the following sufficient condition for stability robustness, which involves the product (PC), is obtained:

$$\frac{\underline{\sigma}[I + PC]\underline{\sigma}[P]}{\bar{\sigma}[PC]} > \bar{\sigma}[\Delta P] \tag{59}$$

Most of the literature on the LQG/LTR method uses the multiplicative uncertainty formulation. An LFSS, however, has inherently parallel structure; therefore, the additive uncertainty formulation is the more natural one. (An additive uncertainty can always be converted to a multiplicative uncertainty by noting that: $L_p = I + \Delta P.P^{-1}$ (assuming that P is square) and that $\bar{\sigma}[L_p - I] \leq \bar{\sigma}[P^{-1}]\bar{\sigma}[\Delta P]$. However, this formulation was found to give very conservative results in preliminary studies. Therefore, it was decided to abandon the multiplicative uncertainty formulation in favor of the additive uncertainty formulation.)

The next step in the design procedure is to design the "target feedback loop", that is, a full state feedback compensator having desirable singular value frequency response. The performance of the closed-loop system depends on the low frequency gain and the crossover frequency of the loop transfer matrix PC; that is, on the behavior of $\underline{\sigma}[PC]$. Larger low frequency gain and crossover frequency indicates better tracking performance. Thus $\underline{\sigma}[PC]$ should lie above the performance specification as shown in Figure 7a. The other requirement is the stability robustness in the presence of model uncertainties. If the multiplicative uncertainty formulation is used, according to (52), the $\bar{\sigma}[PC]$ plot should pass under the robustness barrier $\bar{\sigma}^{-1}(L_p)$ at high frequencies (Figure 7a). On the other hand, if the additive formulation is used, the robustness condition (59) should be satisfied. The target loop (G_{KF}) is designed to satisfy the performance and robutness conditions imposed on the loop gain (PC). The advantage of an LQG-based target loop design is that it has desirable classical properties, and its frequency response can be shaped in the desired manner by varying the weighting

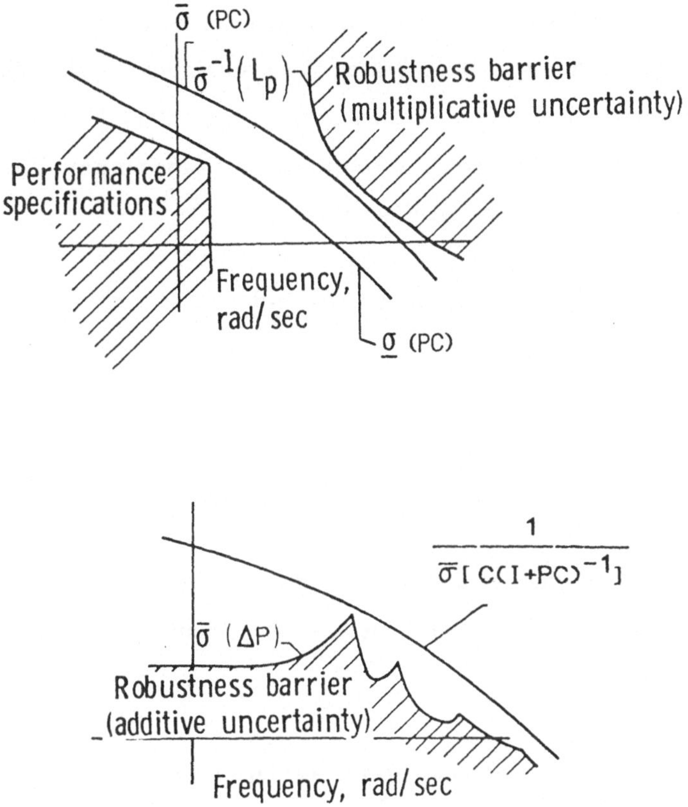

Figure 7. Performance and robustness barriers

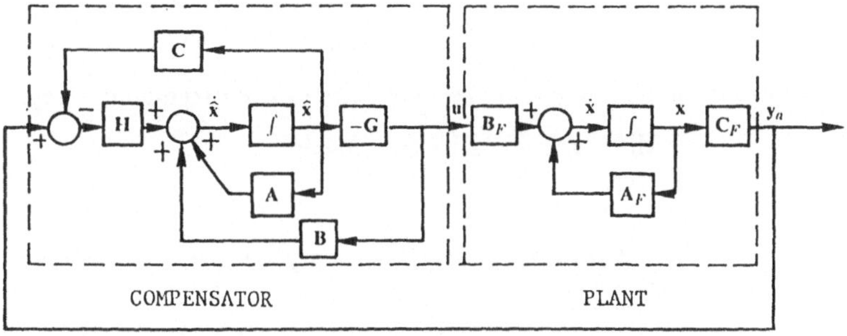

Figure 8. Block diagram of model-based compensator

matrices [Ath.86]. As discussed in [Doy.81], this design can be accomplished using the LQR Riccati equation if the loop is broken at the plant input, or the KBF Riccati equation if it is broken at the point where the residual signal enters the KBF. Herein we select the latter because the objective is to control the attitude output. This selection is also consistent with [Ath.86]. The KBF equations are:

$$A\Sigma + \Sigma A^T + LL^T - \frac{\Sigma C^T C \Sigma}{\mu} = 0 \tag{60}$$

$$H = \Sigma C^T / \mu \tag{61}$$

where L and μ are the design parameters, L being an $n \times m$ matrix, and μ a scalar. H is the KBF gain matrix, and Σ is the corresponding Riccati matrix. The KBF loop transfer matrix is given by:

$$G_{KF}(s) = C(sI - A)^{-1}H \tag{62}$$

Generally, the frequency response $\sigma[G_{KF}(j\omega)]$ would shift higher as μ decreases, and the crossover frequency can be adjusted by changing L [Ath.86].

Having obtained satisfactory singular value behavior of the target loop (i.e., KBF), the third step is to design an LQR to "recover" the desired frequency response. This is accomplished by solving the algebraic Riccati equation

$$A^T P + PA - PBB^T P + \bar{q} C^T C = 0 \tag{63}$$

where P is the Riccati matrix and $\bar{q}$ is a positive scalar. It has been proven in [Doy.81] that the loop transfer matrix $P(j\omega)C(j\omega)$ for the overall system (consisting of the plant, the KBF, and the LQR) converges to $G_{KF}(j\omega)$ pointwise in ω, as $\bar{q} \to \infty$, provided that the open-loop plant has no transmission zeros in the right half plane. The compensator $C(s)$ is given by:

$$C(s) = G(sI - A + BG + HC)^{-1}H \tag{64}$$

where

$$G = B^T P$$

A block diagram of the compensator is shown in Figure 8.

Since the compensation obtained has no guaranteed robustness properties, the last step consists of testing condition (54). It is also desirable to compute the eigenvalues of the entire closed-loop system to ensure the desired degree of stability. The overall closed-loop system is:

$$\begin{bmatrix} \dot{x}_F \\ \dot{\hat{x}}_d \end{bmatrix} = \begin{bmatrix} A_F & -B_F G \\ HC_F & A_d - B_d G - HC_d \end{bmatrix} \begin{bmatrix} x_F \\ \hat{x}_d \end{bmatrix} \tag{65}$$

where the subscript F is used to denote the full-order nominal plant, and $\hat{x}_d$ denotes the state estimate for the design model. If instability is discovered, it will be necessary to return to step 2 and redesign the KBF for lower bandwidth and the LQR for robustness recovery. If this does not produce satisfactory results, it would then be necessary to return to step 1 and include more elastic modes in the design model. Application of the foregoing LQG/LTR procedure to the hoop/column antenna is described in the following section.

3.2.2 Application to Hoop/Column Antenna

Before applying the LQG/LTR method, it is desirable to check the controllability, observability and invariant zero properties of the plant. For the hoop/column antenna, these properties were studied with respect to different sensor/actuator locations. In this design we assume that only one three-axis torque actuator and attitude/rate sensor are used. A measure of the relative controllability of the rigid-body and elastic modes can be obtained by examining the reciprocal condition numbers of the polynomial matrix [Kai.80]

$$C_p(s) = [(sI - A_F), B_F] \tag{66}$$

over the set of eigenvalues of A_F, where A_F and B_F denote the system and input matrices for the full plant (i.e., the 26th order evaluation model). The reciprocal condition number is defined as the ratio of smallest to largest singular values of $C_p(s)$. Comparison of these numbers evaluated at $s = 0$ for the rigid modes and at the eigenvalues of A_F gives an indication of the relative ease or difficulty in affecting the corresponding elastic mode. A similar approach can be used to measure relative observability properties. Based on these analyses, the first three elastic modes have better controllability and observability properties from location 1, which was therefore selected for the actuators and sensors.

The presence of invariant zeros [Ema.82] in the model can be tested by finding the values of s for which

$$\operatorname{rank}\begin{bmatrix} sI - A_F & B_F \\ C_F & 0 \end{bmatrix} < n+3 \tag{67}$$

The finite invariant zeros of the plant are minimum-phase, but just marginally.

The LQG/LTR procedure was subsequently applied, first using only attitude sensors, and then using attitude and attitude-rate sensors. The latter case gave results very similar to the former one. Therefore, only the attitude feedback case is discussed here.

LQG/LTR designs are performed using the design model consisting of

a) only rigid-body model ($n = 6$, $n_q = 0$)

b) rigid-body and the first flexible mode ($n = 8$, $n_q = 1$)

c) rigid-body and the first three flexible modes ($n = 12, n_q = 3$)

The sensor measurements are the three attitude angles at location 1. One, 3-axis torque actuator is used at the same location. The compensator is designed based on these sensors and actuators. The results of the designs are presented in Figures 9-20.

The largest and the smallest singular values of the rigid-body transfer matrix ($n = 6$) are plotted in Figure 9. The corresponding additive uncertainty ΔP, which consists of the (20th order) flexible dynamics, is plotted in Figure 10. Figure 9 clearly shows the roll-off behavior of the plant, which is of the order of $1/s^2$. Figure 10 indicates the presence of poles near the flexible mode frequencies of 0.75 rad/sec, 1.35 rad/sec, etc. Also, the pole near the first mode frequency near 0.75 rad/sec produces the highest peak. It can also be seen that a zero is present near 5 rad/sec. At this stage an "uncertainty envelope," reflecting the uncertainty in the uncontrolled mode parameters, can be drawn as shown in Figure 10. The height of the uncertainty envelope near the peaks of $\bar{\sigma}(\Delta P)$ depends on the "worst case" inherent damping ratios, and the spread around the peaks reflects the uncertainty in the structural mode frequencies. The low-frequency magnitude of the envelope reflects the uncertainty in mode-slopes at actuator and sensor locations. Such uncertainty envelopes would be essentially based on the designer's experience and confidence in the particular mathematical model being used. Therefore we do not show uncertainty envelopes in the subsequent figures, but rather focus on presenting the design methodology.

Based on this rigid design model, the following choice of L and μ were made after a number of trials.

$$L = \begin{bmatrix} 0_3 \\ 10^{-2} I_3 \end{bmatrix}, \mu = 1 \tag{68}$$

The objective is to obtain a bandwidth of at least 0.1 rad/sec. The resulting values of the target loop gain matrix (G_{KF}) are shown in Figure 11. The standard LQG/LTR procedure requires the definition of the target loop transfer characteristics G_{KF} to satisfy the low-frequency performance specifications, and the high-frequency robustness specifications. Thus, in the presence of additive uncertainty ΔP, the procedure states that the robustness condition

$$\frac{\underline{\sigma}[I + G_{KF}]\underline{\sigma}[P]}{\bar{\sigma}[G_{KF}]} > \bar{\sigma}[\Delta P] \tag{69}$$

should be satisfied. However, in the computations performed, it was found that the above condition makes the target loop (G_{KF}) extremely conservative in the sense that higher loop gains can be used without causing instability. Therefore, recovering this conservative loop gain yields a compensator with poor performance. This fact led to a

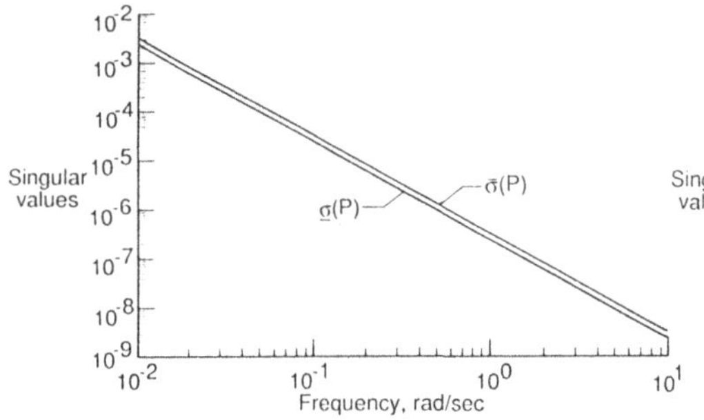

Figure 9. Frequency response for 6th order design model

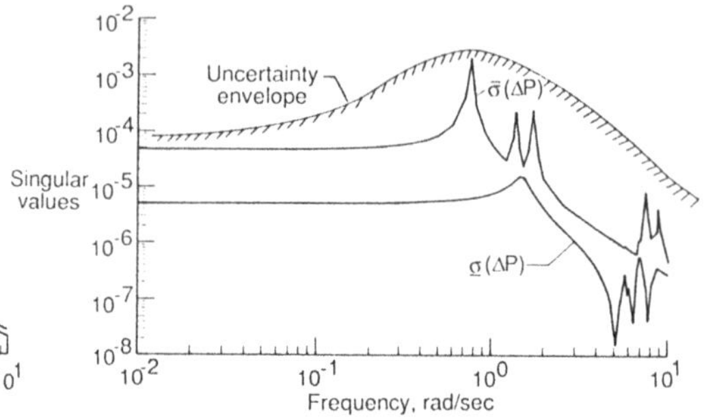

Figure 10. ΔP for 6th order design model

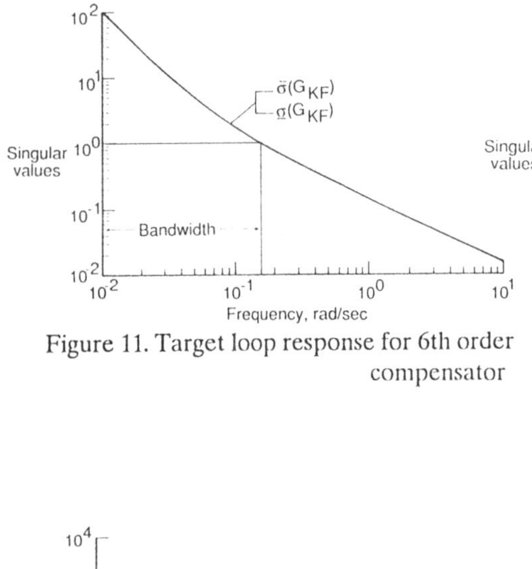

Figure 11. Target loop response for 6th order compensator

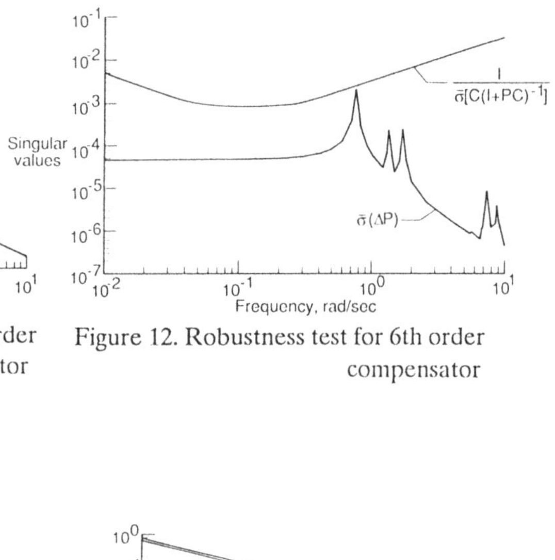

Figure 12. Robustness test for 6th order compensator

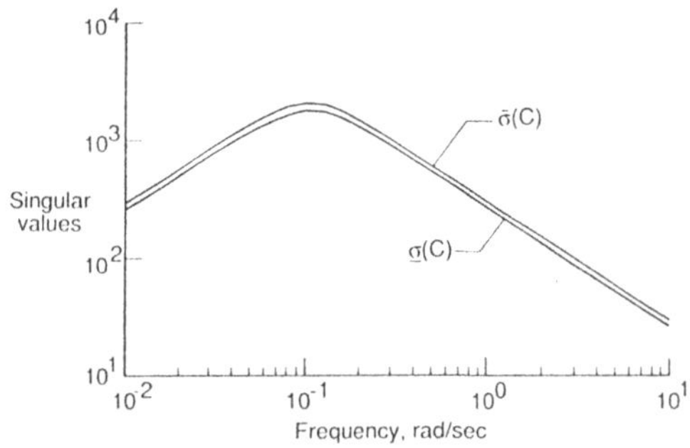

Figure 13. Compensator frequency response (6th order)

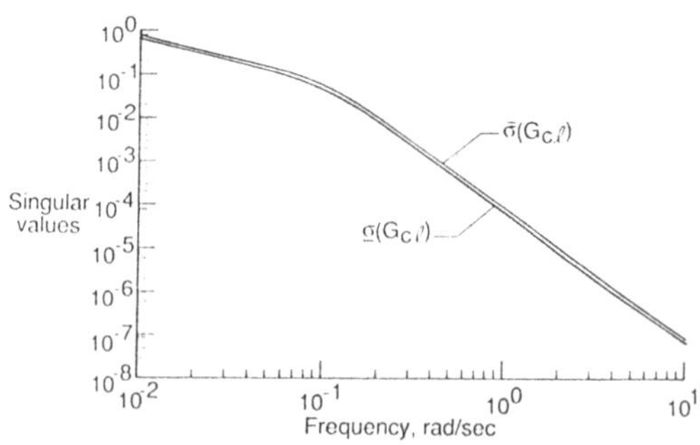

Figure 14. Closed-loop freq. response (6th order compensator)

modification of the LQG/LTR procedure, wherein the target loop is designed to satisfy only the performance specification, and the above robustness test on G_{KF} is omitted. Instead, the recovery is carried out first, and then the (less conservative) stability test (54) is applied directly for the compensator C.

Using the recovery procedure, the compensator is obtained for this case with $\bar{q} = 10^4$. The resulting stability test (Eq. 54) is shown in Figure 12. It is seen that the stability margin is lowest at the first mode frequency (0.75 rad/sec). Any increase in the gain (obtained by making $\bar{q} > 10^4$) resulted in the violation of the stability condition (54). The resulting singular values of the compensator C are given in Figure 13, which shows that the compensator is of "lead-lag" type. The overall loop bandwidth is obtained from the singular values of the closed-loop transfer function $G_{c\ell}$ shown in Figure 14. It is seen that the bandwidth (i.e., the frequency at which $\underline{\sigma}(G_{c\ell})$ starts to roll off below unity) is far short of the required 0.1 rad/sec. Thus it became evident that it is not possible to meet the performance specifications using the rigid-body design model.

To overcome the above problems, the inclusion of the first flexible mode (0.75 rad/sec) in the design model was considered. The inclusion of the first flexible mode, which is predominantly a torsion mode, results in a design model of order 8. However, the best compensator obtained with this design model did not satisfy the bandwidth specifications.

The next step was to additionally include the second and third flexible modes simultaneously in the design model. It is logical to do this because they represent the first bending modes about the X and Y axes. Thus the 3 flexible modes included in the design model were: the first torsion mode and the first bending modes in the XZ and YZ planes. The order of this design model was 12. The singular value plots for P and ΔP are shown in Figures 15 and 16 respectively. Figure 15 indicates that that P has zeros near 0.082 and 0.22 rad/sec, and poles near 0.75, 1.35, and 1.7 rad/sec. It is seen from the ΔP plot (Figure 16) that $\bar{\sigma}(\Delta P)$ is considerably lower than that in Figure 10. The target loop gain (G_{KF}) obtained after several trials is shown in Figure 17. The recovery was accomplished with $\bar{q} = 10^{10}$. The stability test is shown

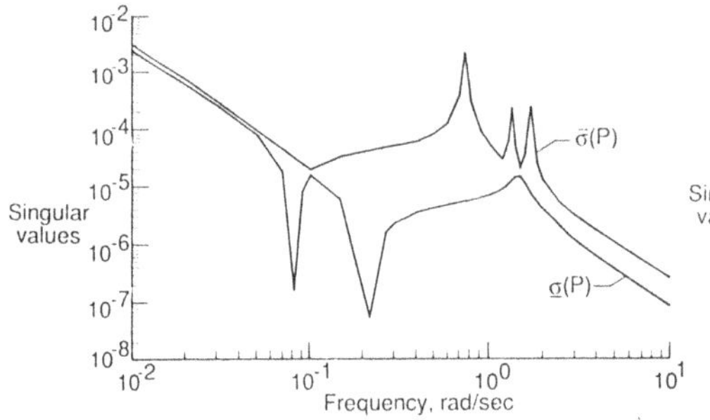

Figure 15. Frequency response for 12th order design model

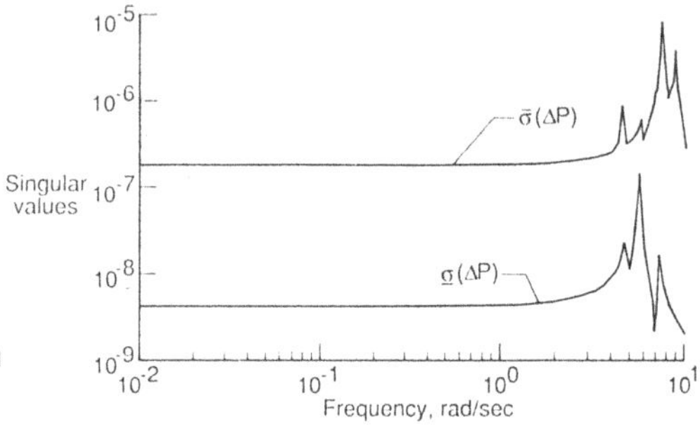

Figure 16. ΔP for 12th order design model

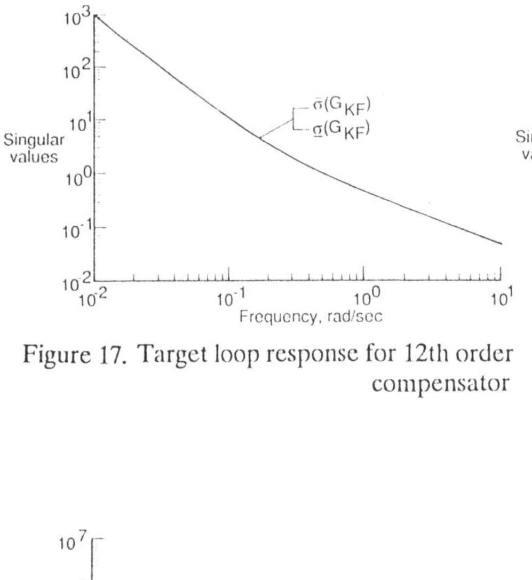

Figure 17. Target loop response for 12th order compensator

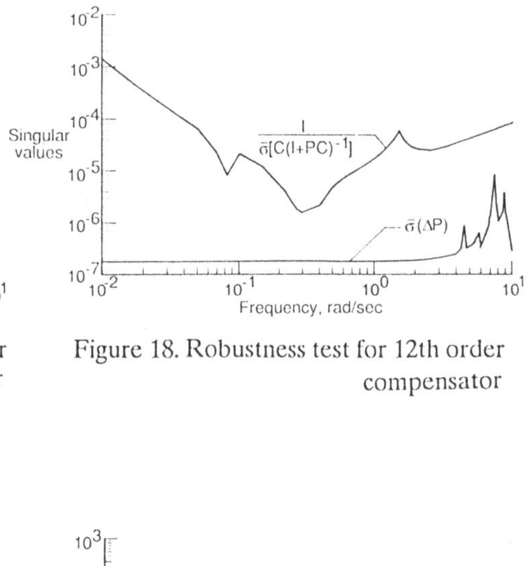

Figure 18. Robustness test for 12th order compensator

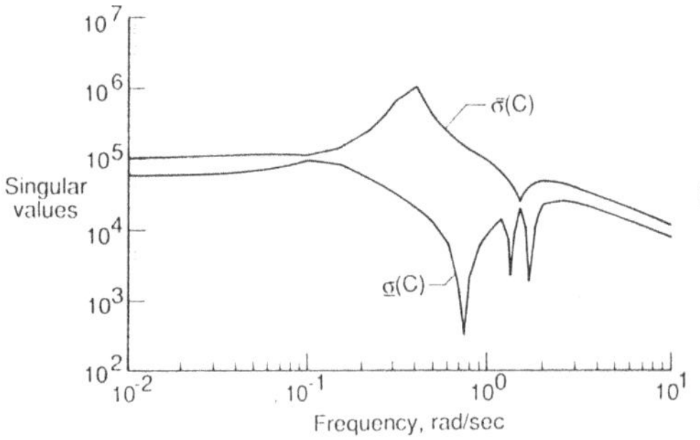

Figure 19. Compensator frequency response (12th order)

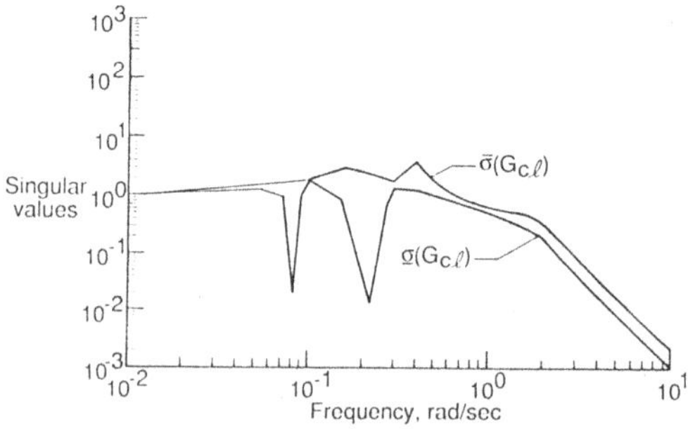

Figure 20. Closed-loop freq. response (12th order compensator)

in Figure 18, wherein condition (54) is satisfied with a wide margin. Also, at the peaks for ΔP (near 8 rad/sec) the upper curve slopes upward, indicating good tolerance of high-frequency uncertainty. Generally, the LQG/LTR technique attempts to choose C in such a way that P is approximately inverted and the product PC is replaced by G_{KF}. The design model has eigenvalues at -0.0075 $\pm$ j0.75, -0.0135 $\pm$ j1.035, and -0.017 $\pm$ j1.7. Figure 19 shows that C has zeros with frequencies near these locations. The design model also has transmission zeros at $-0.9 \times 10^{-4} \pm j0.082, -0.37 \times 10^{-3} \pm j0.22$, and $-0.29 \times 10^{-3} \pm j0.22$, which are close to the $j\omega$ axis, and tend to numerically behave as nonminimum phase. Some alleviation is obtained by the compensator pole near 0.4 rad/sec. The plots for the the closed-loop transfer function $G_{c\ell}$ are given in Figure 20. It is seen that a bandwidth of 0.1 rad/sec is obtained except for the presence of a transmission zero near 0.082 rad/sec which causes some deterioration of performance. This zero is invariant under feedback, and depends on the sensor/actuator locations.

The above results indicate that the LQG/LTR design can provide good robutness to unmodeled dynamics while meeting the performance specifications. However, it turns out that the controller is very sensitive to inaccuracies in the design model.† In particular, a small error (of only about 2%) in the design model elastic-mode frequencies causes a large increase in $\bar{\sigma}(\Delta P)$ and results in the violation of the stability condition (54)! In other words, when applied to the LFSS problem, the method can handle high-frequency uncertainties very well, but cannot handle low-frequency uncertainties. This can be attributed to the following facts: 1) the controller is an "inversion" controller which attempts pole-zero cancellation, and 2) the natural damping is very small, causing the peaks in $\bar{\sigma}(\Delta P)$ to be very sharp. These findings indicate that the LQG/LTR method is not suitable for LFSS control unless the design model frequencies are known with very high accuracy, which is rarely the case. (It should be noted, however, that the parameter identifiability of natural frequenceies is very high, as will be discussed in Chapter 4). To a certain extent, high sensitivity to errors in the design model frequencies appears to be also a characteristic of other compensators which use the design model in the "prediction" part of the observer. One solution to this problem is to use time-domain

---

†The author is indebted to Prof. Arthur E. Bryson for his incisive analysis and comments on this problem.

LQG designs employing many actuators and sensors, which seem to have considerably lower sensitivity to errors in the design model. Another possible method for alleviating this problem is to use a damping enhancement controller (such as the CDEC, Chapter 2) in the inner loop to increase the damping and reduce the peaks of $P(s)$. However, this would result in coupling all the elastic modes, and neither the additive nor the multiplicative uncertainty formulation would be applicable. Reduction of sensitivity to inaccuracies in the design model frequencies, while using the fewest possible actuators and sensors, remains an open area of research.

As for the transmission zeros within the performance bandwidth, it may be possible to shift them by changing actuator and sensor locations. Alternatively they can be eliminated altogether if more actuators are used, because non-square plants generally have no transmission zeros. In any case, it is important to give a carefuly consideration to the locations and the numbers of actuators and sensors right from early design phases of an LFSS.

3.3 Effect Of Actuator Nonlinearities

In the previous sections we investigated the design of LQG-type controllers for LFSS, which are robust to unmodeled dynamics and parameter errors. Robustness to modeling errors is only a part of the overall robustness requirement. In addition, the controller must also be robust to actuator and sensor nonlinearities and failures. In that spirit, we investigated the robustness of different types of dissipative controllers to actuator and sensor nonlinearities in Chapter 2. In this section, we investigate the robustness of LQG-type controllers to actuator and sensor nonlinearities.

The LQG-based controller design, whether time or frequency domain, is essentially a linear design. Although the assumption of linearity of the model itself is justified in a neighborhood of the steady-state, the actuators and sensors available in practice almost always have nonlinear characteristics. These characteristics include, but are not limited to, saturation, dead-zones and hysteresis. Our investigation of robustness to nonlinearities utilizes the fact that the full state feedback, infinite duration LQ regulator (LQR) has been shown to have robustness to actuator nonlinearities belonging to the $(0.5,\infty)$ sector [And.71; Saf.77]. However, most realistic nonlinearities do not lie entirely

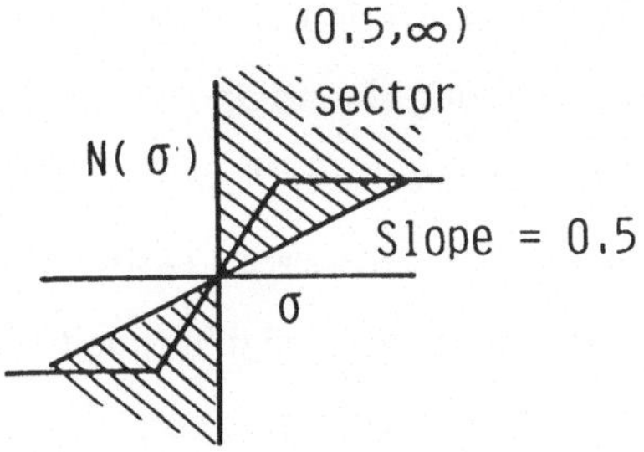

Figure 21. Type I nonlinearity (saturation)

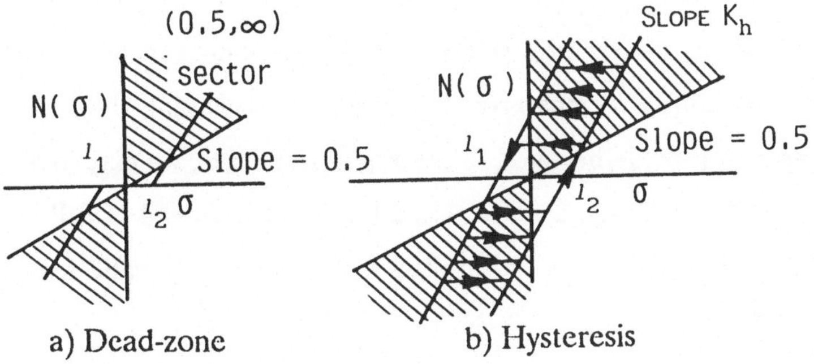

Figure 22. Type II nonlinearities

in that sector. For example, the saturation nonlinearity (Figure 21) satisfies the sector condition in a finite region containing the origin ($\sigma = 0$), but violates it in regions away from the origin. We call such nonlinearities "Type-I" nonlinearities. On the other hand, nonlinearities such as dead zone and hysteresis satisfy the sector condition in regions away from the origin, but violate it in a neighborhood of the origin (Figure 22). We term such characteristics "Type-II" nonlinearities.

We first consider the robustness of LQR (i.e., full state feedback) to actuator nonlinearities. In the next sub-section, we investigate Type-I actuator nonlinearities, and prove that there exists a region of attraction in the state space such that all trajectories starting in that region converge exponentially to the origin. We then prove that, when Type-II nonlinearities are present, there exists a region of ultimate boundedness in the state space such that all trajectories starting outside the region will enter the region in a finite time T, and will remain in the region for $t \geq T$. Our results utilize the fact that the LQR Riccati matrix provides a natural Lyapunov function. The expressions obtained for the stability regions provide methods for selecting the LQR performance function weights to provide better stability regions. Finally, we extend these results to LQG-type controllers wherein a state estimator is used instead of full state feedback. The results presented herein are not limited to LFSS, but are valid for general linear, time-invariant systems.

3.3.1 Type-I Nonlinearities: Region of Attraction

Infinite-duration, linear quadratic regulators (LQR) have highly desirable robustness properties, which include tolerance to certain types of nonlinearities at the input [And.71; Saf.77]. In particular, the closed-loop system is globally asymptotically stable (g.a.s.) if the nonlinearities in the control channels lie in the (0.5,∞) sector; i.e., $N_i(0) = 0$ and $\sigma N_i(\sigma) > 0.5\sigma^2$ for $\sigma \neq 0$ (for the ith nonlinearity). As stated previously, Type-I nonlinearities such as saturating amplifiers (Figure 21) violate the (0.5,∞) sector away from the origin ($\sigma = 0$). In this section, we extend the LQR robustness results to a broader class of nonlinearities which satisfy the sector condition at least in a neighborhood of the origin. An estimate of the region of attraction is obtained using a method similar to [Wal.67] and special properties of the Riccati equation. We consider a general

LTI system given by:

$$\dot{x} = Ax + Bu \tag{70}$$

where x, u are n and m dimensional state and control vectors, and A, B are $n \times n$ and $n \times m$ constant matrices. Consider the infinite duration regulator problem where the following performance function is minimized:

$$J = \int_0^\infty e^{2\alpha t}(x^T Qx + u^T Ru)dt \tag{71}$$

where $\alpha \geq 0$ is a scalar which represents the required degree of stability, Q is an $n \times n$ symmetric positive semidefinite matrix, and R is an $m \times m$ positive definite matrix. The control $u(t)$ which minimizes J in Equation (71) is given by: $u = Gx$, where

$$G = -R^{-1}B^T P \tag{72}$$

$$A^T P + PA + 2\alpha P + Q - PBR^{-1}B^T P = 0 \tag{73}$$

Assuming that (A, B) is controllable and $(Q^{1/2}, A)$ is observable, the Riccati equation (73) has a unique positive definite solution P, and the eigenvalues of $(A + BG)$ have real parts $< -\alpha$. Defining $V(x) = x^T Px$, it can be proved [And.71] that $V(x(t)) \leq e^{-2\alpha t}V(x(0))$. That is, the system has the degree of stability α.

Considering the case where nonlinearities exist in the control actuators, the commanded control input is: $u_c = Gx$, but the actual input is:

$$u = N(u_c) = [N_1(u_{c1}), N_2(u_{c2}), ..., N_m(u_{cm})]^T \tag{74}$$

where $N_i(\nu)$ denotes (a possibly time-varying) nonlinear gain function. Define an m-vector valued function ϕ such that:

$$\phi(\sigma) = N(\sigma) - 0.5\sigma \tag{75}$$

where σ denotes the $m \times 1$ argument.

Region of Attraction

Consider the set $\Sigma_a \subset R^m$ such that $\sigma^T R\phi(\sigma) > 0$ for $\sigma \,\epsilon\, \Sigma_a$. Let S_1 denote the inverse image of Σ_a, i.e., $S_1 = \{x | Gx \,\epsilon\, \Sigma_a\}$, and let ∂S_1 denote the boundary of S_1. The following theorem gives an estimate of the region of attraction.

Theorem 3. *If the condition : $\sigma^T R\phi(\sigma) \geq 0$ is satisfied for $\sigma \,\epsilon \Sigma_a$, the closed-loop system is asymptotically stable (a.s.), and S_a is a region of attraction, where $S_a = \{x | x^T Px < d\}$, and $d = \min\limits_{x \epsilon \partial S_1} x^T Px$. Furthermore, the system has the degree of stability α inside S_a.*

Outline of Proof. Considering the function $V(x) = x^T Px$, it can be shown using Eqs. (72), (73) and (74) that

$$\dot{V}(x) = -x^T(Q + 2\alpha P)x - 2u_c^T R\phi(u_c) \tag{76}$$

Since Σ_a contains a neighborhood of the origin of R^m, S_1 contains a neighborhood of the origin of R^n. The largest region $R_\delta = \{x | V(x) \leq \delta\}$ which is a subset of S_1 can be shown to be S_a. The proof follows since $V > 0$ and $\dot{V} \leq -2\alpha V$ along all trajectories in S_a. ∎

As an important special case suppose the ith input channel has nonlinearity $N_i(\sigma_i)$, which lies in the $(0.5, \infty)$ sector for $\sigma_i \,\epsilon\, [\ell_{1i}, \ell_{2i}]$, $(i = 1.2, .., m)$, where $\ell_{1i} < 0$ and $\ell_{2i} > 0$. For this case,

$$S_1 = \cap_{i=1}^m L_i, \tag{77}$$

where

$$L_i = \{x | g_i^T x \epsilon [\ell_{1i}, \ell_{2i}]\} \tag{78}$$

(g_i^T denotes the ith row of G). That is, L_i is bounded by two hyperplanes: $g_i^T x = \ell_{1i}$, and $g_i^T x = \ell_{2i}$. The following corollary is an immediate cosequence of Theorem 1.

Corollary 3.1. *Suppose R is diagonal with entries R_{ii}, $i \epsilon [1, m]$ and the ith nonlinearity lies in the $(0.5, \infty)$ sector for $\sigma_i \epsilon [\ell_{1i}, \ell_{2i}]$. Then an estimate of the region of attraction is given by S_a, where*

$$S_a = \{x | x^T P x < d\} \tag{79}$$

$$d = \min_{\substack{i\epsilon[1,m] \\ j\epsilon[1,2]}} (\ell_{ji} R_{ii})^2 / b_i^T P b_i \tag{80}$$

where b_i denotes the ith column of B. The system has the degree of stability α inside S_a. ■

Corollary 3.1 enables one to readily determine an estimate of the region of attraction for an LQ design, given ℓ_{ji}. Furthermore, a method is provided for adjusting weights to get a larger region of attraction. An analytical explanation of the dependence of S_a on R and α can be obtained by letting $R = \rho \hat{R}$, where $\hat{R} > 0$ is a diagonal matrix, and $\rho > 0$ is a scalar. Using (79) and (80), S_a is given by:

$$x^T P_0 x < \min_{\substack{i\epsilon[1,m] \\ j\epsilon[1,2]}} (\ell_{ji} \hat{R}_{ii})^2 / b_i^T P_0 b_i \tag{81}$$

where $P_0 = P/\rho$. From (73),

$$(A + \alpha I)^T P_0 + P_0 (A + \alpha I) - P_0 B \hat{R}^{-1} B^T P_0 = -Q/\rho \tag{82}$$

Taking partial derivative of (82) w.r.t. ρ and rearranging,

$$A_c^T\left(-\frac{\partial P_0}{\partial \rho}\right)+\left(-\frac{\partial P_0}{\partial \rho}\right)A_c=-Q/\rho^2\leq 0 \tag{83}$$

where $A_c=(A+\alpha I-B\hat{R}^{-1}B^TP_0)$. Because A_c is strictly Hurwitz, $\dfrac{\partial P_0}{\partial \rho}\leq 0$, and P_0 is a non-increasing function of ρ. [That is, $P_0(\rho_2)-P_0(\rho_1)\leq 0$ (negative semidefinite) if $\rho_2>\rho_1$]. Thus the region S_a of (81) increases (not necessarily strictly) as ρ increases. Also, if $|\ell_{1k}|$ or ℓ_{2k} is small compared to the $|\ell_{1i}|$ and ℓ_{2i} for the other nonlinearities (i.e., if the kth nonlinearity violates the sector condition closer to the origin than the other nonlinearities), one may increase the weight R_{kk} to increase "d" in (82), which gives a larger region of attraction. Next, for a fixed ρ, taking partial derivative of (73) w.r.t. α, it can be shown (in a similar manner) that $\dfrac{\partial P}{\partial \alpha}>0$, which implies from (79) and (80) that S_a decreases as α increases.

An important question that arises is: how large can the region of attraction be made by increasing R? The following theorem provides the answer.

Theorem 4. *Asymptotic Properties: Under the conditions of Corollary 1, suppose B is of rank m, and $R=\rho\hat{R}$, where $\rho>0$ is a scalar and $\hat{R}>0$ is a diagonal matrix. Then*

a) *If Re$\{\lambda_i(A)]\}\leq-\alpha$ (i=1,2,..,n), the region S_a can be made arbitrarily large by increasing ρ.*

b) *If Re $\{\lambda_j(A)\}>-\alpha$ for some j, then the region S_a tends to a constant, bounded or semi-bounded region as $\rho\to\infty$.*

Proof. The Riccati equation (73) becomes

$$(A+\alpha I)^TP+P(A+\alpha I)-\rho^{-1}PB\hat{R}^{-1}B^TP=-Q \tag{84}$$

a) Suppose Re$[\lambda_i(A)]\leq-\alpha$ (i=1,2,...,n). That is, $(A+\alpha I)$ is a Hurwitz matrix. The limiting properties of the Riccati matrix (as $\rho\to\infty$) are different when $(A+\alpha\, I)$ is merely *Hurwitz* and when it is *strictly Hurwitz*. Therefore, we consider these two subcases separately. First considering the subcase when $(A + \alpha I)$ is strictly Hurwitz, $\lim_{\rho\to\infty} P = \bar{P} > 0$ exists [Jos.84b]. From (79) and (80), the limiting S_a is given by

$$x^T \bar{P} x < \min_{\substack{i\epsilon[1,m] \\ j\epsilon[1,2]}} \rho^2 \ell_{ji}^2 \hat{R}_{ii} / b_i^T \bar{P} b_i \tag{85}$$

Thus S_a can be made arbitrarily large by increasing ρ.

Now consider the second subcase where one or more eigenvalues of $(A+\alpha I)$ lie *on the imaginary axis*. In this subcase, $\lim_{\rho\to\infty} P$ generally does not exist. It was shown previously that $P_0 = P/\rho$ is monotonically non-increasing and bounded from below $(P_0 > 0)$ (regardless of whether $(A+\alpha I)$ is Hurwitz or not). Therefore, $\lim_{\rho\to\infty} P_0 = \bar{P}_0 \geq 0$ always exists. For the present subcase, we will prove by contradiction that $\bar{P}_0 = 0$. Suppose $\bar{P}_0 \neq 0$. Letting $\hat{V}(t) = z^T(t)\bar{P}_0 z(t)$, where $z(t)$ satisfies: $\dot{z} = (A + \alpha I)z$, choose a $z(0)$ such that $\hat{V}(0) > \nu$ for any given $\nu > 0$. Considering (82) with the right-hand side equal to zero in the limit, it can be proved that $\dot{\hat{V}}(t) = z^T(t)\bar{P}_0 B\hat{R}^{-1}B^T\bar{P}_0 z(t) \geq 0$. Thus $\hat{V}(t)$ is a non-decreasing function of t. It is also bounded because $\mathrm{Re}\{\lambda_i(A+\alpha I)\} \leq 0$. Therefore, $\lim_{t\to\infty} \hat{V}(t)$ exists, and $\dot{\hat{V}} \to 0$ as $t \to \infty$. Since B is of full rank, $B\hat{R}^{-1}B^T > 0$, and $\bar{P}_0 z(t) \to 0$; therefore, $z^T(t)\bar{P}_0 z(t) = \hat{V}(t) \to 0$. But $\hat{V}(t)$ is non-decreasing, which implies $V(t) \geq V(0) > \nu$; therefore, our assumption that $\bar{P}_0 \neq 0$ is incorrect, and $\bar{P}_0$ must be equal to zero. Substituting $P_0 = P/\rho$ and $R = \rho\hat{R}$, S_a is given by (81). From (81), since $P_0 \to 0$ as $\rho \to \infty$, S_a can be made arbitrarily large by increasing ρ.

b) As proved in part (a), $\lim_{\rho\to\infty} P_0 = \bar{P}_0 \geq 0$ always exists. Suppose $\bar{P}_0 = 0$. Then $G \to 0$, which implies $(A + \alpha I + BG)$ will have unstable eigenvalues in the limit, which is not true [Kwa.72b]. Therefore, $\bar{P}_0 \neq 0$, and S_a tends (in the limit) to the following region, which is bounded in one or more directions in R^n :

$$x^T \bar{P}_0 x < \bar{d} = \min_{\substack{i\epsilon[1,m] \\ j\epsilon[1,2]}} \ell_{ji}^2 \hat{R}_{ii}^2 / b_i^T \bar{P}_0 b_i \tag{86}$$

∎

Example: Application to Hoop/Column Antenna

We consider the problem of controlling the attitude of the 122m hoop/column antenna about the Y-axis; i.e., the pitch-axis attitude. The attitude angle, of course, includes the rigid and elastic motion. The linearized pitch-axis model is given by:

$$J\ddot{\theta} = T_1 + T_2 \tag{87}$$

$$\ddot{q}_i + 2\rho_i\omega_i\dot{q}_i + \omega_i^2 q_i = \Psi_{1i}T_1 + \Psi_{2i}T_2 \tag{88}$$

where J is the Y-axis moment of inertia, θ is the rigid-body pitch angle, and T_1 and T_2 are the Y-axis control torques applied by control- moment-gyros (CMG's) at points 1 and 2 shown in Figure 5, Chapter 2. We consider only the first two bending modes about Y-axis, i.e., modes 2 and 7 in the complete hoop/column antenna model given in Chapter 1. Thus q_i, ρ_i, ω_i denote the modal amplitude, inherent damping ratio, and the natural frequency for the ith mode, and Ψ_{ji} denotes the ith mode-slope at actuator location j. The state vector can be defined as: $x = (\theta, \dot{\theta}, q_1, \dot{q}_1, q_2, \dot{q}_2)^T$.

Case 1: Two saturating actuators with same limit. Suppose both of the actuators saturate for $|T_c| > T_{\max}$, where T_c and $T_{\max}$ denote the command and saturation torques respectively. The $(0.5, \infty)$ sector is violated for $|T_c| \geq 2T_{\max}$. A nominal LQ regulator design was first performed (using diagonal Q and R and with α =0) to obtain a rigid-body closed loop frequency and damping ratio of 0.138 rad/sec and 0.707 respectively, and closed-loop structural mode damping ratios of at least 0.5. Two measures of the size of S_a were considered. The first measure was the maximum value of the angular displacement (rigid-body + elastic) at one of the sensor locations (e.g. at actuator no. 1), within the hyperellipsoid S_a, which is given by $Y_{\max} = \sqrt{c^T P^{-1} cd}$ where c^T is the corresponding (1 $\times$ n) output matrix (i.e., $y = c^T x$). Another measure of the size of S_a is its volume V', which is proportional to $1/\sqrt{\det(P)}$. In the example considered, $Y_{\max}$ and V' showed similar behavior for changes in the design parameters. Because it is physically more meaningful, $Y_{\max}$ was used as a measure of S_a.

For the nominal LQ design, $Y_{\max}$ was 68.5 degrees. Starting with 1 percent of its

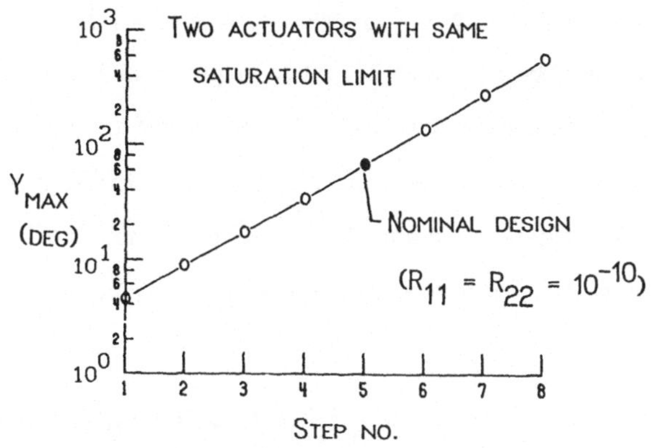

Figure 23. Effect of increasing R

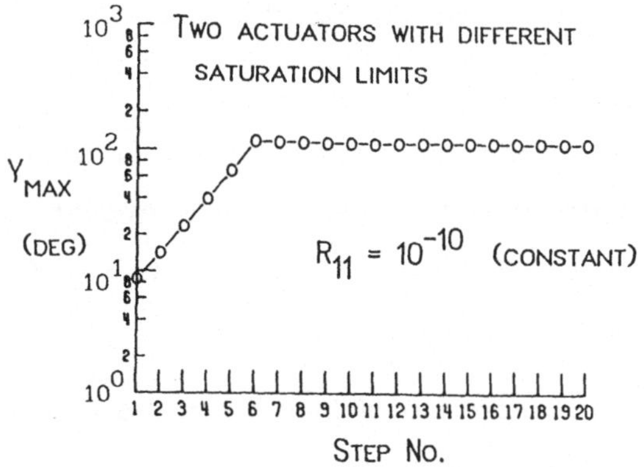

Figure 24. Effect of increasing R_{22}

nominal value, R (i.e., both R_{11} and R_{22}) was increased by a factor of $\sqrt{10}$ at each step. $Y_{\max}$ increases as R increases (Figure 23) since larger R causes smaller control efforts, which are less likely to reach the saturation limits. However, increase in R would also cause the nominal performance to deteriorate. According to Theorem 4, since A has no eigenvalues in the open right-half plane, S_a can be made arbitrarily large by increasing R. When α was increased from zero in steps of 0.02, $Y_{\max}$ decreased as analytically predicted.

Case 2: Two saturating actuators with different limits. Suppose the saturation limits for actuators 1 and 2 are $T_{\max}$ and 0.125 $T_{\max}$ respectively. For this case, the nominal LQ design yielded $Y_{\max}$ of only, 8.6°. As discussed previously, increasing only weight R_{22} on the input for actuator 2 (by a factor of 2 at each step) while keeping R_{11} constant resulted in increased $Y_{\max}$ (Figure 24). Beyond the sixth step, however, $Y_{\max}$ flattens out because R_{11} (which is constant) becomes the dominant factor which determines d [see (80)].

In general, Lyapunov methods are known to give conservative results. Thus the estimate of the region of attraction obtained above is expected to be conservative. In our limited simulation studies, the estimated region of attraction ($Y_{\max}$) was about 6-10 times smaller than that obtained by simulation.

We consider stability in the presence of Type-II nonlinearities in the following subsection.

3.3.2 Type II Nonlinearities: Region of Ultimate Boundedness

In this subsection we consider the closed-loop stability of LTI systems controlled by infinite duration LQR, when Type-II actuator nonlinearities are present in the loop. We prove that there exists a region of ultimate boundedness such that all trajectories will enter that region in a finite time and will remain in that region. Therefore, limit cycles, if any, will stay inside that region. The method used is similar to that in [Wei.68], and also utilizes the special structure of the algebraic Riccati equation.

The system under consideration is given by eq. (70), and the problem is to minimize the performance function in (71). In addition to the assumptions in the previous section, we require that α is positive if Q is singular. Because of the nonlinearities that exist in the control actuators, the commanded control input is u_c, but the actual input is $u = N(u_c)$, as in (74).

Region of Ultimate Boundedness

Let $\Sigma_b \subset R^m$ denote a bounded region containing the origin. Suppose the nonlinearity $N(\sigma)$ is such that $\sigma^T R\phi(\sigma) \geq 0$ outside Σ_b, that is, only in the complement Σ_b^c of Σ_b. Let S_2 denote the inverse image of Σ_b in R^n, i.e., $S_2 = \{x \mid Gx \,\epsilon\, \Sigma_b\}$, and let ∂S_2 denote the boundary of S_2. The following theorem gives an estimate of the region of ultimate boundedness.

Theorem 5. *If the condition: $\sigma^T R\phi(\sigma) \geq 0$ is satisfied in Σ_b^c, and if $\phi(\sigma)$ is bounded in Σ_b, the region S_b is a region of ultimate boundedness for the closed-loop system, where*

$$S_b = \{x \mid x^T P x \leq h\} \tag{89}$$

$$h = \max_{\Omega_1 \cup \Omega_2} x^T P x \tag{90}$$

$$\Omega_1 = \{x \mid x^T \hat{Q} x = \mu,\; x \,\epsilon\, S_2\} \tag{91}$$

$$\Omega_2 = \{x \mid x^T \hat{Q} x < \mu,\; x \,\epsilon\, \partial S_2\} \tag{92}$$

$$\hat{Q} = Q + 2\alpha P \tag{93}$$

$$\mu = -2 \min_{\sigma \epsilon \Sigma_b} \sigma^T R\phi(\sigma) \tag{94}$$

Outline of Proof. Starting with the Lyapunov function $V(x) = x^T P x$ where P is the Riccati matrix, it can be shown that

$$\dot{V} = -x^T \hat{Q} x - 2u_c^T R\phi(u_c) \leq -x^T \hat{Q} x + \mu \tag{95}$$

where $\hat{Q}$ and μ are defined in Eqs. (93) and (94). Thus $\dot{V} < 0$ if $u_c^T R\phi(u_c) \geq 0$ (*i.e.*, $x \in S_2^c$), or if $x \in S_3$, where

$$S_3 = \{x \mid x^T \hat{Q} x > \mu\} \tag{96}$$

Therefore, $\dot{V} < 0$ for $x \in S_3 \cup S_2^c$. The trajectories are ultimately bounded in a region S_b of the form (89), if $V > 0$ and $\dot{V} < 0$ outside that region, and if $V(x) \to \infty$ as $\| x \| \to \infty$. To get the least conservative estimate of the region of ultimate boundedness, it is necessary to obtain the smallest hyperellipsoid (of the form (89)) containing the region $(S_3 \cup S_2^c)^c = S_2 \cap S_3^c$. The smallest hyperellipsoid containing a region is the one containing its boundary, which, in this case is $\Omega_1 \cup \Omega_2$. ■

Figure 25 shows the regions S_2^c , S_3, boundaries Ω_1, Ω_2, etc. for a two-dimensional case. As shown, the region S_b is then the boundary and the interior of the ellipsoid: $x^T Px = h$, which is the smallest ellipsoid enclosing the unshaded region containing the origin. (i.e., h is obtained by minimizing $x^T Px$ over $\Omega_1 \cup \Omega_2$). Except in simple cases (such as second-order systems with Ω_2 described by straight lines), analytical solutions cannot be obtained, and numerical algorithms for constrained optimization must be used. Since our objective is to analytically investigate the relationships between the performance function weights and the region of ultimate boundedness, we seek a closed-form expression to approximate h. One such approximation would be to obtain the smallest hyperellipsoid enclosing: $x^T \hat{Q} x = \mu$. This is given by: $x^T Px = h'$, where

$$h' = \max_{x^T \hat{Q} x = \mu} x^T Px \tag{97}$$

As shown in Figure 25, this hyperellipsoid encloses the one given by: $x^T Px = h$, and is therefore more conservative.

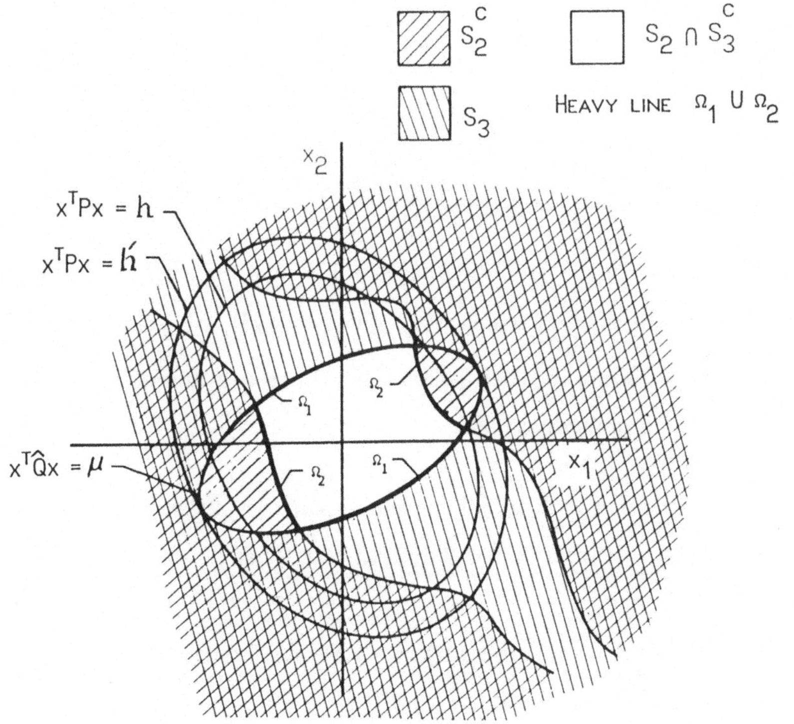

Figure 25. Estimation of region of ultimate boundedness

Theorem 6. *Under the conditions of Theorem 5, an estimate of the region of ultimate boundedness for the closed-loop system is given by*

$$S_b' = \{x \mid x^T P x \leq h'\} \tag{98}$$

$$h' = \mu / [2\alpha + \lambda_m(P^{-1}Q)] \tag{99}$$

where λ_m denotes the smallest eigenvalue.

Proof. Proceeding as in the proof of Theorem 5, $\dot{V} < 0$ for $x^T \hat{Q} x > \mu$ (i.e., for $x \in S_3$). The smallest hyperellipsoid enclosing S_3^c is given by (98), where $h' = \max\limits_{x^T \hat{Q} x = \mu} x^T P x = \mu \lambda_M(\hat{Q}^{-1} P)$. Further simplification yields (99). Since $V > 0$ and $\dot{V} < 0$ in $[S_b']^c$, S_b' is a region of ultimate boundedness. ■

For the case where R is diagonal, suppose $\sigma_i \phi_i(\sigma_i) \leq 0$(i.e., the sector condition is violated) for $\sigma_i \in [\ell_{1i}, \ell_{2i}]$ where $\ell_{1i} < 0$ and $\ell_{2i} > 0$. Then

$$\mu = -2 \min_{\sigma \in \Sigma_b} \sum_{i=1}^{m} R_{ii} \sigma_i \phi_i(\sigma_i) = -2 \sum_{i=1}^{m} R_{ii} \min_{\sigma_i \in [\ell_{1i}, \ell_{2i}]} \{\sigma_i \phi_i(\sigma_i)\} \tag{100}$$

Since $\sigma_j \phi_j(\sigma_j) < 0$ for some $j \in [1, m]$, $\mu > 0$. For making S_b or S_b' small, μ should be made as small as possible. The above expression suggests that μ can be made smaller by reducing the weights R_{kk} corresponding to those input channels which have nonlinearities which most severely violate the $(0.5, \infty)$ sector. (i.e., for $\sigma_k \phi_k(\sigma_k)$ most negative in the violation region). Also, (99) suggests that h' can be made smaller by a) increasing α and b) decreasing R (or increasing Q). The dependence of S_b' on R and α can be analytically investigated as follows. Let $R = \rho \hat{R}$, where $\hat{R} > 0$ is a diagonal matrix and $\alpha > 0$ is a scalar. From (98)-(100), denoting $P_0 = P/\rho$, S_b' is given by:

$$x^T P_0 x \leq M / \{2\alpha + \lambda_m(P^{-1}Q)\} \tag{101}$$

where

$$M = -2 \min_{\sigma \epsilon \Sigma_b} \sum_{i=1}^{m} \hat{R}_{ii} \sigma_i \phi_i(\sigma_i) = -2 \sum_{i=1}^{m} \hat{R}_{ii} \min_{\sigma_i \epsilon [\ell_{1i}, \ell_{2i}]} \{\sigma_i \phi_i(\sigma_i)\}$$

It was shown in Sec. 3.3.1 that $\partial P_0/\partial p \leq 0$, that is, P_0 increases as ρ decreases. It can also be shown similarly that, as ρ decreases, P decreases; that is, P^{-1} increases. Therefore, the region S_b' given by (101) becomes smaller (not necessarily strictly) as ρ decreases. It can also be proved in a similar manner that, for a fixed ρ, P is a strictly increasing function of α, which implies [from (98) and (99)] that S_b' becomes smaller as α increases (provided that $\{2\alpha + \lambda_m(P^{-1}Q)\}$ increases. This will certainly be the case when Q is singular).

An important question that arises is: how small can S_b' be made by changing ρ and α? Since P is a strictly increasing function of α, S_b' can be made arbitrarily small by increasing α [see (98), (99)]. However, the gain G can become unreasonably large. For fixed α, when $\rho \downarrow 0$, it is not possible to draw general conclusions about the behavior of S_b' without making additional assumptions (such as minimum-phase plant). The following theorem addresses a special case.

Theorem 7. *Under the conditions of Theorem 6, suppose $R = \rho\hat{R}$, where $\hat{R} > 0$, diagonal, and $\rho > 0$ is a scalar. Suppose $Q > 0$, $m = n$, and B is nonsingular. Then S_b' can be made arbitrarily small by decreasing ρ.*

Proof. The Riccati equation is:

$$(A + \alpha I)^T P + P(A + \alpha I) - \frac{PB\hat{R}^{-1}B^T P}{\rho} = -Q \tag{102}$$

Following [Kwa.72b], $\lim_{\rho \downarrow 0}[PB/\sqrt{\rho}] = P_1$ exists. Since $m = n$ and B is nonsingular, $\lim_{\rho \downarrow 0}[P/\sqrt{\rho}] = P_2 \geq 0$ exists. Taking the limit of (102) as $\rho \downarrow 0$, $P_2 B\hat{R}^{-1}B^T P_2 = Q$, which implies $P_2 > 0$. Thus $P_0 = P/\rho \approx P_2/\sqrt{\rho}$ tends to ∞ as $\rho \downarrow 0$. Therefore, from (101), S_b' can be made arbitrarily small by decreasing ρ. ■

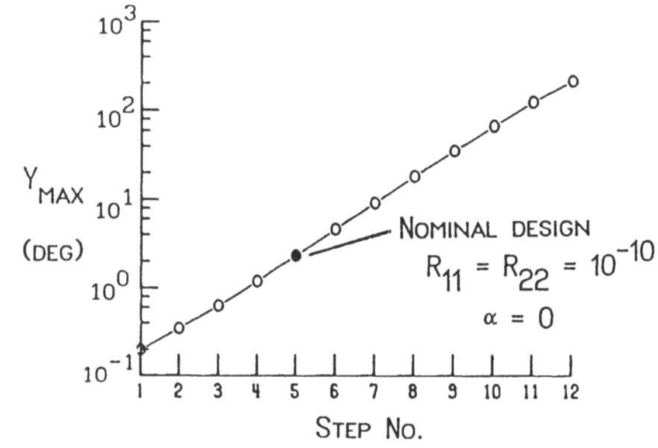

Figure 26. Effect of increasing R

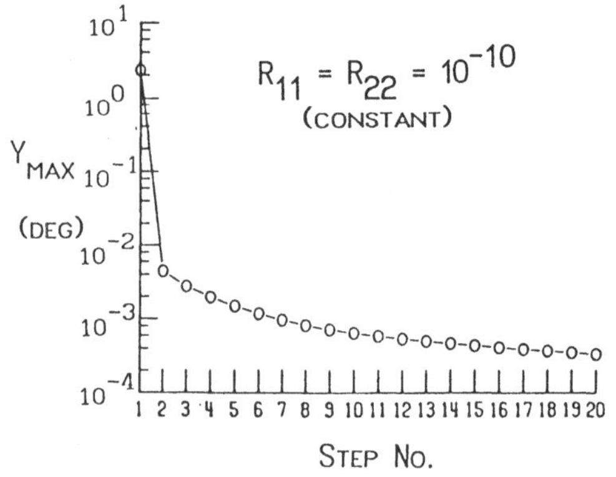

Figure 27. Effect of incraesing α

Example: Application to Hoop/Column Antenna

For the pitch-axis attitude control problem considered in Sec. 3.3.1, suppose each of the two torque actuator has hysteresis as shown in Figure 22b, with $K_h = 1$ and a=0.25 N-m. The region of ultimate boundedness, S_b', is given by (98) and (99). The smaller the size of S_b', the better is the tolerance of the design to these nonlinearities. As in Sec. 3.3.1, the maximum possible angular displacement Y_{max} within the region of ultimate boundedness S_b' is used as a measure of the size of S_b'. ($Y_{max} = \sqrt{(c^T P^{-1} c h')}$ where c^T is the 1×6 output matrix).

The nominal LQ design as described in sec. 3.3.1 (with α =0) yielded $Y_{max} = 2.32$ degrees for the region of ultimate boundedness. Starting with one percent of its nominal value, R was increased by a factor of $\sqrt{10}$ at each step. Consistent with the analysis presented, this caused $Y_{\max}$ to increase (Figure 26). When α was increased from zero in steps of 0.02, $Y_{\max}$ showed a large decrease from the first to the second step, and slower decrease after that (Figure 27). The large decrease was because $\{2\alpha + \lambda_m(P^{-1}Q)\}$ appears in the denominator in (99), and α is large compared to $\lambda_m(P^{-1}Q)$. Based on the analytical and the numerical results, α appears to be an important design parameter which can significantly affect the estimate of the region of ultimate boundedness. Selecting α to be a small positive scalar rather than zero can significantly reduce the size of S_b'. Perhaps this fact should be taken into account while designing LQG-based compensators for LFSS, using both time-domain and frequency-domain approaches.

3.3.3 Extension to State-Estimate Feedback

So far we have investigated the stability of systems with full state feedback LQ control law, when actuators have Type-I or Type-II nonlinearities. We now extend these results to the more practical case when a state estimator (observer) is in the loop. The problem with using state estimators is that the standard LQR robustness properties [including tolerance to (0.5,∞) sector nonlinearities] are lost [Doy.78]. However, it has been proven in [Saf.78, Vid.80] that if a control law using state vector feedback stabilizes a linear or nonlinear system, then the same control law using feedback of non-divergent state estimates will also stabilize it. In this sub-section, we consider two types of non-divergent state estimators, namely, an exponentially stable state estimator, and an estimator with ultimately bounded estimation error. We then obtain expressions for the regions of attraction and ultimate boundedness.

Suppose an $(n \times 1)$ estimate $\hat{x}$ of the state vector is used to generate the control input. That is,

$$u_c = G\hat{x} \tag{103}$$

where G is given by (72) and the state estimator or detector is given by:

$$\dot{\hat{x}} = f(\hat{x}, y, u_c, t) \tag{104}$$

In (104), $y = h(x,t)$ denotes the $\ell \times 1$ sensor output vector, and f is a continuous function of its arguments. Let $\tilde{x} = x - \hat{x}$ denote the $n \times 1$ state estimation error vector. We consider two classes of state estimators based on the following assumptions:

(A1) Suppose there exist positive definite symmetric matrices $\tilde{P}$, $\tilde{Q}$ such that for all $\tilde{x} \in R^n$,

$$\tilde{V}(\tilde{x}) = \tilde{x}^T \tilde{P} \tilde{x} \tag{105}$$

$$\dot{\tilde{V}}(\tilde{x}) \leq -\tilde{x}^T \tilde{Q} \tilde{x} \tag{106}$$

Or

(A2) There exist $\tilde{P}, \tilde{Q} > 0$, and a scalar $\tilde{\mu} > 0$ such that for $\tilde{V}$ defined by (105),

$$\dot{\tilde{V}} \leq -\tilde{x}^T \tilde{Q} \tilde{x} + \tilde{\mu} \tag{107}$$

If assumption (A1) is satisfied, then the estimator is globally exponentially stable, whereas if (A2) is satisfied, the estimation error is ultimately bounded. The assumption

(A1) would indeed be satisfied if the estimation error dynamics is linear and exponentially stable. This situation can arise when using an extended Kalman filter, if the actuator and sensor nonlinearities are known, and the sensor nonlinearities have slopes $\geq 1/2$. (See [Saf.78, Jos.87]). The assumption (A2) would be satisfied when sensor nonlinearities possibly have slopes $< 1/2$ in bounded regions containing the origin. (e.g., dead-zones). Because it is physically more meaningful, we assume R to be diagonal. There are m control channels, and the ith channel has a nonlinear actuator gain $N_i (i = 1, 2, , m)$. The actual input u is given as in eq. (74). We assume that each N_i is Lipschitz-continuous; that is, there exist $K_i < \infty$ such that

$$| N_i(\sigma_1) - N_i(\sigma_2) | \; < \; K_i \,| \sigma_1 - \sigma_2 | \text{ for } i = 1, 2, .., m \tag{108}$$

Let

$$K = diag(K_1, K_2, ..., K_m) \tag{109}$$

Stability Regions

We first consider the case when Type I nonlinearities are present in the actuators.

Theorem 8. *Suppose the state estimator satisfies the assumption (A1), and all N_i are of Type I. Then the closed-loop system given by Eqs. (70), (74), (103), (104) is exponentially stable, and S_a is a region of attraction, where*

$$S_a = \{x, \tilde{x} \mid x^T P x + \delta \tilde{x}^T \tilde{P} \tilde{x} < \min_{\substack{i \epsilon [\ell, m] \\ j \epsilon [1,2]}} (\ell_{ji} R_{ii})^2 / b_i^T P b_i\} \tag{110}$$

where

$$\delta = \lambda_M(G^T K R G) / \{\lambda_m(Q + 2\alpha P)\lambda_m(\tilde{Q})\} \tag{111}$$

[$\lambda_M(.)$ *and* $\lambda_m(.)$ *denote the largest and the smallest eigenvalue respectively*].

Outline of Proof. Letting

$$V(x,\tilde{x}) = x^T P x + \tilde{x}^T \tilde{P} \tilde{x} \tag{112}$$

we have

$$\dot{V} = 2x^T P \dot{x} + 2\tilde{x}^T \tilde{P} \dot{\tilde{x}} \tag{113}$$

Using (72), (73), (103), (106), (108) in (113), along with the fact that: $\sigma[N_i(\sigma) - 0.5\sigma] \geq 0$ for $\sigma \in [\ell_{1i}, \ell_{2i}]$, simplifying (by making use of the algebraic Riccati equation), we have

$$\dot{V} \leq -z^T \hat{Q} z - 2u_c^T R[N(u_c) - (1/2)u_c] \tag{114}$$

where

$$z = (x^T_, \tilde{x}^T)^T \tag{115}$$

$$\hat{Q} = \begin{bmatrix} Q + 2\alpha P & -G^T RKG \\ -G^T RKG & \delta \tilde{Q} \end{bmatrix}$$

If δ is chosen as in (111), $\hat{Q}$ can be shown to be positive semidefinite. Using a procedure similar to that in Theorem 3, it can be proved that $V > 0$ and $\dot{V} < 0$ along all trajectories in S_a, and S_a is a region of attraction. ■

Theorem 9. *Suppose the estimator satisfies the assumption (A1), and all $N_i(i = 1, 2, , m)$ are Type II nonlinearities. Under these conditions, for the system given by (70), (74), (103) and (104), the state $x(t)$ is ultimately bounded in the region S_b given by:*

$$S_b = \{x \mid x^T P x \leq h\} \tag{117}$$

where

$$h = \mu / \min[\{\lambda_m(P^{-1}Q) + 2\alpha\}, \lambda_m\{\tilde{P}^{-1}\tilde{Q}\}] \tag{118}$$

(min[.,.] denotes the smaller of the two arguments).

$$\mu = -2 \min_{u_c} \sum_{i=1}^{m} R_{ii} u_{ci} [N_i(u_{ci}) - (1/2) u_{ci}] \tag{119}$$

Outline of Proof. Starting with V as in (112), we obtain (114). Thus $V > 0$ and $\dot{V} < 0$ for $z^T \hat{Q} z > \mu$. Therefore, an estimate of the region of ultimate boundedness for the closed-loop system is given by: $\{z \mid z^T P z \leq \mu \lambda_M(\hat{Q}^{-1}\hat{P})\}$ where $\hat{P} = diag(P, \tilde{P})$ (See Theorem 6). Since

$$\lambda_M(\hat{Q}^{-1}\hat{P}) = \lambda_m^{-1}(\hat{P}^{-1}\hat{Q}) = 1/\lambda_m \begin{bmatrix} P^{-1}Q + 2\alpha I & -P^{-1}G^T RKG \\ -\delta^{-1}\tilde{P}^{-1}G^T RG & \tilde{P}^{-1}\tilde{Q} \end{bmatrix} \tag{120}$$

for sufficiently large δ, the region of ultimate boundedness is given by

$$x^T P x + \delta \tilde{x}^T \tilde{P} \tilde{x} \leq h \tag{121}$$

where h is as in (118). Thus $x(t)$ is ultimately bounded in S_b. ∎

The region S_b differs from the region S_b' for the full state feedback case (Theorem 6) because the estimator-dependent term $\lambda_m(\tilde{P}^{-1}\tilde{Q})$ is present in (118). This term is a measure of the speed of response of the estimator. Larger $\tilde{P}^{-1}\tilde{Q}$ implies larger $|\dot{\tilde{V}}/\tilde{V}|$, which implies a faster estimator. Thus, for sufficiently fast estimator, S_b would be the same as the region S_b' for the full state feedback case of Theorem 6.

When the estimation error does not go to zero, but is ultimately bounded, the system with Type I nonlinearities is (generally) not asymptotically stable. However, when Type II nonlinearities are present, it is proved below that the $x(t)$ is ultimately bounded.

Theorem 10. *Suppose the state estimator satisfies assumption (A2), and all N_i (i=1,2,...,m) are of Type II. Then $x(t)$ is ultimately bounded in $\bar{S}_b$, where*

$$\bar{S}_b = \{x \mid x^T P x \leq \bar{h}\} \tag{122}$$

$$\bar{h} = (\mu + \tilde{\mu}) / \min[\{\lambda_m(P^{-1}Q) + 2\alpha\}, \lambda_m(\tilde{P}^{-1}\tilde{Q})\}] \tag{123}$$

Proof. The proof is very similar to that of Theorem 9. ■

Some Remarks About Stability Regions

In this sub-section, we investigated the closed-loop stability of linear multivariable systems controlled by LQG-type controllers when the actuator nonlinearities escape the (0.5, ∞) stability sector. The cases considered included state estimators which are globally exponentially stable, or with ultimately bounded estimation error. Expressions were obtained for the regions of attraction and ultimate boundedness for different types of nonlinearities and state estimators. The research in this area is as yet incomplete. Future efforts should include more detailed analyses of state estimators with the desired properties. Some results on that topic were obtained in [Jos.87], but further research is needed. In addition, it would be highly desirable to reduce the conservativeness of the estimates of stability regions (perhaps along the lines of [Kar.75]). Finally, we need to incorporate these results in the controller synthesis procedure to design a compensator which is robust to modeling errors as well as actuator/sensor nonlinearities.

Chapter 4

Related Topics

In the previous chapters we investigated techniques for designing robust controllers for LFSS. The starting point of controller design is a mathematical model, usually a finite element model of the LFSS. Such models tend to be inaccurate beyond the first three or four elastic modes; hence the need for robust controllers. The larger the modeling errors, the lower is the performance of robust controllers designed to incorporate them. Therefore it is important to have as accurate parameter estimates as possible. In this chapter, we first investigate the identifiability of LFSS parameters, which include natural frequencies, damping ratios, and mode shapes or slopes. This is accomplished by obtaining the Cramér-Rao lower bound on the covariance of the parameter estimation error, which indicates the best achievable estimation accuracy. Numerical results of identifiability studies are presented using the hoop/column antenna model. We also briefly discuss another important topic in LFSS control, namely large-angle maneuvering. This addresses the problem of rotating the LFSS to a new attitude in order to point to a new target. Such maneuvers must be accomplished in the least possible time, with minimum expenditure of energy, and without exciting the elastic modes. Finally, areas for future research in control of LFSS are discussed.

4.1 Parameter Identifiability Studies

Control systems synthsis techniques, especially those based on the LQG theory, require a reasonably good knowledge of the design model as well as the uncontrolled modes (if only to get an upper bound on the contribution of the latter). Determination of the parameters, (e.g., using finite element methods) generally does not give sufficiently accurate parameter estimates. This is particularly true for damping ratios, because finite element methods have no means of computing inherent damping ratios. In fact there appear to be no rigorous proven methods for modeling inherent structural damping. The inherent damping ratios play a vital role in the controller design process. For example, the magnitudes of the peaks and notches in the singular value frequency response are determined by the inherent damping ratios. These magnitudes often determine the critical points in singular value robustness tests (see Chapter 3).

Ground testing for parameter estimation is almost impossible for two reasons: 1) many of the LFSS will be assembled in space, and 2) deployable structures, 100 m or larger in size, would be too large to test on ground. They are not designed to withstand stresses of the 1- g environment (i.e., gravity). In addition, because of the sheer size, it would be impractcal to construct the necessary test facilities such as vaccum chambers.

In view of these problems, the best strategy appears to be as follows:

1) Design an interim dissipative controller based on the available *a priori* knowledge of the parameters. Such controllers are known to have excellent robustness to parameter errors, actuator/sensor nonlinearities, and failures. They can maintain at least the stability, and keep the vibration amplitudes (and the accompanying stresses in the LFSS) sufficiently small, during deployment, assembly, and initial operating phase.

2) Estimate the LFSS parameters on orbit while it is being controlled by the interim dissipative controller. Design a high-performance controller (e.g. an LQG-type MBC) using these parameter estimates, and implement it. This controller would use the same actuators and sensors, so that the switch-over from the interim controller is accomplished by only re-programming the control computer.

Thus it is important to have an effective technique for on-orbit parameter estimation. Several methods have been developed for parameter estimation over the last three decades. These include: least squares methods, minimum-variance estimators, maximum likelihood estimators, extended Kalman filtering, stochastic approximation method, etc. (See [Sor.80], [Sag.71]). Herein we shall not discuss any specific techniques for identification, but rather how well can we identify various LFSS parameters using the best of these methods. That is, we investigate the thoretical lower limits on the achievable estimation errors.

4.1.1 Cramér-Rao Lower Bounds

Since most established parameter estimation methods use discrete-time formulation, we consider the discretized LFSS model expressed as:

$$x(k+1) = Fx(k) + Gu(k) + v(k) \tag{1}$$

$$y(k) = Cx(k) \tag{2a}$$

$$z(k) = y(k) + w(k) \tag{2b}$$

where x, u, y represent $n \times 1$, $m{\times}1$ and $\ell \times 1$ state, control, and output vectors, v, w represent zero-mean, mutulally uncorrelated white noise processes, and z is the observation (or sensor output) vector. F, G, and C are appropriately dimensioned constant matrices. If $x(0), v(k), w(k)$ are Gaussian, the conditional density $f(Z_N/\theta)$ is Gaussian, where

$$Z_N = \{z(0), z(1), ..., z(N)\} \tag{3}$$

and θ is the $p \times 1$ vector of parameters to be estimated. The matrices F, G, and C are functions of θ. Assuming $v = 0$, we have [Sor.80]

$$f(Z_N/\theta) = C_1 \exp\left[-\frac{1}{2}\sum_{k=0}^{N}\{z(k) - y(k)\}^T W^{-1}\{z(k) - y(k)\}\right] \tag{4}$$

where C_1 is constant and W is the covariance matrix of $w(k)$. If the a-priori density of $\theta, f(\theta)$ is Gaussian, the unconditional density $f(Z_N)$ can be shown to be also Gaussian.

Suppose $\hat{\theta}$ is an absolutely unbiased estimate of θ based on Z_N. Then it can be proved that [Sor.80]

$$\mathcal{E}[(\theta-\hat{\theta})(\theta-\hat{\theta})^T/\theta] \geq J^{-1} \tag{5}$$

where $\mathcal{E}$ denotes the expected value, and

$$J = -\mathcal{E}\left\{\frac{\partial^2}{\partial\theta^2}\ln f(Z_N/\theta)/\theta\right\} \tag{6}$$

Using (4) and (6), it can be shown that

$$J = \sum_{k=0}^{N}\left\{\frac{\partial y(k)}{\partial\theta}\right\}^T W^{-1}\left\{\frac{\partial y(k)}{\partial\theta}\right\} \tag{7}$$

J is called the Fisher information matrix. From (5), J^{-1} is a lower bound on the covariance of the parameter estimation error vector. It is also *the best lower bound* under the stated assumptions (i.e., equality can hold in (5) for some absolutely unbiased estimate). J^{-1} is called the "Cramér-Rao lower bound".

Suppose θ is a scalar. Then the sensitivity propagation equations can be written as:

$$\begin{bmatrix} x \\ \frac{\partial x}{\partial\theta} \end{bmatrix}_{k+1} = \begin{bmatrix} F & 0 \\ \frac{\partial F}{\partial\theta} & F \end{bmatrix}\begin{bmatrix} x \\ \frac{\partial x}{\partial\theta} \end{bmatrix}_k + \begin{bmatrix} G \\ \frac{\partial G}{\partial\theta} \end{bmatrix} u(k) \tag{8}$$

and the output snsitivity is:

$$\frac{\partial y_{(k)}}{\partial\theta} = \frac{\partial C}{\partial\theta}x(k) + C\frac{\partial x_{(k)}}{\partial\theta} \tag{9}$$

Similar equations can be obtained for the case where θ is a vector. If the system is asymptotically stable, J in (7) tends to a constant as $N \to \infty$, if there is no persistent input to the systems. Under these condtions, if the initial state $x(0) = 0$, J would be zero. If persistent input is present, the information keeps growing, and $J \to \infty$ as $N \to \infty$. If all the paremeters are identifiable, J is nonsingular.

Our objective herein is to gain insight into the degree of identifiability of various LFSS parameters, i.e., natural frequencies, damping ratios, mode shapes, and mode slopes. This can be accomplished by comparing the Cramér-Rao standard deviations (i.e., square roots of the diagonal elements of J^{-1}) for these parameters, obtained using zero input and nonzero initial state.

Considerable volume of literature exists on the design of optimal input for parameter identification. If nonzero inputs are used during identification, they would have to be sufficiently small so as to avoid significant excitation of elastic motion, to safeguard the integrity of the LFSS. Also, for accurate identification, precise measurements of the input (with no bias and only small noise) are required. For these reasons, it is reasonable to use zero input, with nonzero initial state, for parameter estimation. Nonzero initial state can be obtained by using one or more actuators, which would subsequently be turned off, and the data collected for N samples could then be used in an on-line or off-line identification algorithm.

Sensitivity Equations

For the purpose of identifiability studies for the elastic mode parameters, we consider only the elastic motion. We also assume that only torque actuators and attitude rate sensors (e.g., rate gyros) are used during parameter identification, and that the parameters to be estimated are natural frequencies, damping ratios, and mode slopes (i.e., rotational mode shapes) at the sensor locations. The equation of unforced motion for the ith mode $(i = 1, 2, ..., n_q)$ is given by:

$$\ddot{q}_i + 2\rho_i \omega_i \dot{q}_i + \omega_i^2 q_i = 0 \tag{10}$$

The $l \times 1$ output vector due to elastic motion is given by:

$$y = \sum_{I=1}^{n_q} \begin{bmatrix} \phi_{1i} \\ \phi_{2i} \\ \vdots \\ \phi_{\ell i} \end{bmatrix} q_i \tag{11}$$

where ϕ_{ji} denotes the mode slope of the ith mode at location j. The parameters to be estimated for the ith mode are:

$$\theta_i = [\rho_i, \omega_i, \phi_{1i}, \phi_{2i}, ..., \phi_{\ell i}]^T \tag{12}$$

Thus there are $(\ell + 2)$ parameters per mode, or a total of $n_q \times (\ell + 2)$ parameters. The functional dependence of the continuous-time equations on these parameters is quite simple; therefore, it is convenient to derive the sensitivity equations in continuous time, and then to discretize them. The sensitivity state equations for the ith mode with respect to ρ_i and ω_i are given by:

$$\frac{\partial \ddot{q}_i}{\partial \rho_i} + 2\rho_i\omega_i\left(\frac{\partial \dot{q}_i}{\partial \rho_i}\right) + \omega_i^2\left(\frac{\partial q_i}{\partial \rho_i}\right) + 2\omega_i\dot{q}_i = 0 \tag{13}$$

and

$$\frac{\partial \ddot{q}_i}{\partial \omega_i} + 2\rho_i\omega_i\left(\frac{\partial \dot{q}_i}{\partial \omega_i}\right) + \omega_i^2\left(\frac{\partial q_i}{\partial \omega_i}\right) + 2\rho_i\dot{q}_i + 2\omega_i q_i = 0 \tag{14}$$

With zero input, the sensitivity of q with respect to the mode slopes is zero. The output sensitivity matrix for the ith mode is:

$$\frac{\partial y}{\partial \theta_i} = \left[\frac{\partial y}{\partial \rho_i}, \frac{\partial y}{\partial \omega_i}, \frac{\partial y}{\partial \phi_{1i}}, ..., \frac{\partial y}{\partial \phi_{\ell i}}\right]_{\ell \times (\ell+2)} \tag{15}$$

The output sensitivity terms on the right-hand side of (15) can be obtained form (11) and (15) can be written as:

$$\frac{\partial y}{\partial \theta_i} = [E_{i1}x_{ai}, E_{i2}x_{ai}, ..., E_{i,\ell+2}x_{ai}] \tag{16}$$

where x_{ai} is the augmented state vector for the ith mode:

$$x_{ai} = \left[q_i, \dot{q}_i, \frac{\partial q}{\partial \rho_i}, \frac{\partial \dot{q}_i}{\partial \rho_i}, \frac{\partial q_i}{\partial \omega_i}, \frac{\partial \dot{q}_i}{\partial \omega_i}\right]^T \tag{17}$$

Thus the information matrix J can be computed by propagating the (descretized version of) augmented state vectors (x_{ai}) for all the modes, and using (7) and (16). Note that the initial conditions for all the sensitivity states are zero. The Cramér-Rao bound is then given by J^{-1}.

Numerical Results

Using the 10-elastic-mode finite element model of the 122m hoop/column antenna, Cramér-Rao bounds were computed for various elastic parameters. Nonzero initial state was created using the torquer at actuator location No. 1 (Chapter 2, Fig. 5). Two, three- axis rate gyros were assumed at locations 1 and 2. The parameters to be estimated are: ρ_i, ω_i, $\phi_{x_1 i}$, $\phi_{x_2 i}$, $\phi_{y_1 i}$, $\phi_{y_2 i}$, $\phi_{z_1 i}$, $\phi_{z_2 i}$, where ϕ_{abc} denotes the c'th mode slope in a-direction at location b. A sampling frequency of 4 Hz. (2.5 times the 10th mode frequency), and data length of 130 sec. ($\simeq$ time constant for the first mode), were used in the study. The rate gyro sensor noise was assumed to have a standard deviation intensity of 1 arc-sec/sec. (Cramér-Rao bounds for other noise levels can be obtained simply by multiplying by the standard deviation intensity).

An examination of the magnitudes of the mode-slopes for each mode corresponding to X, Y and Z axes revealed the contributions of each mode to the rate gyro outputs. The modes for which the mode-slope magnitudes are high at a particular gyro location, should be identified using that gyro data for the corresponding axis. Nonzero initial states were attained by using the torquer at actuator location 1 to generate a rectangular wave with nominal amplitude of 10 ft-lbs. and with frquency equal to that of the first mode. Figure 1 shows typical Cramér-Rao standard deviations as they evolve with time. Since the motion for the higher modes decays much faster, Cramér-Rao standard deviations for the higher modes attain their steady-state values relatively quickly. Table 1 shows some examples of steady-state Cramér-Rao standard deviations (σ%) expressed as a percentage of the true value of the parameter.

From the results obtained, it can be concluded that natural frequencies have the best identifiabiity (i.e., the smallest Cramér- Rao lower bounds), followed by damping ratios, and then mode-slopes. The fact that damping ratios have excellent identifiability is particularly welcome because it is usually the worst-known (*a priori*) parameter. The identifiability of mode-slopes varies from fair to very good. The larger the magnitude of the mode-slope, the better is its identifiability. This suggests that poorly identifiable

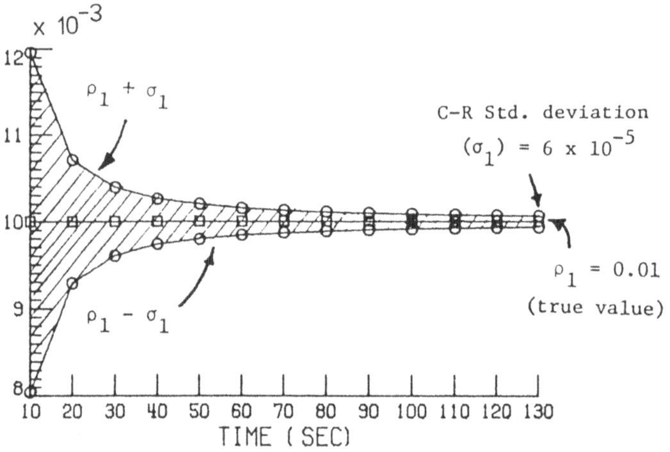

Figure 1. Evolution of Cramér-Rao (C-R) bounds (typical)

Table 1. Cramér-Rao bounds

Mode	ω	σ%	ρ	σ%	$\lvert\phi_1\rvert$	σ%	$\lvert\phi_2\rvert$	σ%
1	0.75	.04	.01	0.65	0.5E-2	0.28	0.16E-4	52
3	1.7	.006	.01	0.87	0.37E-2	0.57	0.37E-3	3.7
10	8.8	.08	.01	12	0.24E-2	15	0.49E-2	8.8

(σ% = C-R Standard deviation expressed as a percentage of the Parameter)

mode-slopes (or shapes) are usually small in magintude and do not contribute to the outputs. Therefore, poorly identifiable mode shapes or slopes can perhaps be ignored (assumed to be zero) for the purpose of controller design. Unlike mode-slopes, the identifiability of damping ratios (ρ) improves as ρ's decrease.

4.2 The Maneuvering Problem

Much of this book has concentrated on the design of control systems for fine-pointing and vibration suppression. This is essentially a linear design problem, except for actuator and sensor nonlinearities. Another important control problem is large-angle attitude-maneuvering of LFSS. For this problem, the dynamic equations are generally nonlinear. Furthermore, the rigid and elastic modes are usually coupled.

Unlike the linear case, there is no generic structure for nonlinear LFSS models. The finite element method is perhaps the most commonly used method for modeling of small elastic motion about a steady state. However, it cannot be used in a straight-forward manner for deriving nonlinear models for arbitrarily rotating and translating LFSS. The basic principles (e.g., Hamilton's principle. {See [Jun.86]}) for deriving hybrid (ordinary/ partial differential equation) models are well-established. An ordinary-differential-equation model can be obtained by incorporating discretization in the basic formulation. Methods such as the assumed modes method can be used for discretization. A model obtained in this manner usually exhibits a high degree of coupling between rigid and elastic motion, as well as between different elastic modes. No general mathematically tractable form of the model exists, and one has to derive models individually for specific LFSS. Examples of derivation of nonlinear LFSS models may be found in [Jun.86], [Kak.87] and [Mei. 87]. As in the linear case, the question of how many modes should be included in the discretization still exists. Unfortunately there appear to be no systematic order reduction methods for nonlinear systems.

After deciding on the number of elastic modes to be included, the equations of motion take the form (ignoring noise):

$$\dot{x} = f(x, u, t) \tag{18}$$

$$y = g(x, u, t) \tag{19}$$

where

$$x = (\varsigma^T, v^T, \alpha^T, \omega^T, q^T, \dot{q}^T)^T \tag{20}$$

In (20), ς, v denote the 3×1 inertial position and velocity vectors of the rigid-body center of mass (c.m.); α, ω denote the 3×1 rigid-body attitude (Euler) angle vector and the 3×1 rigid-body angular velocity vector; q is the $n_q \times 1$ elastic modal amplitude vector. The output vector y consists of attitude and rate measurements at various sensor locations.

4.2.1 Optimal Control Problem Formulation

The optimal large-angle attitude maneuvering problem can be formulated as that of obtaining the control input $u(t)$ and the final time t_f which minimize the performance index:

$$J = W \cdot t_f + x^T(t_f)Sx(t_f) + \int_0^{t_f} (x^T Qx + u^T Ru)dt \tag{21}$$

$$W > 0, \quad S = S^T \geq 0, \quad Q = Q^T \geq 0, \quad R = R^T > 0$$

subject to (18), and the constraints:

$$x(0) = x_0; \quad \psi[x(t_f), t_f] = 0 \tag{22}$$

$$p[x(t), u(t), t] \geq 0 \tag{23}$$

where ψ and p are some vector functions. For example, for a rest-to-rest maneuver, $\psi[x(t_f), t_f] = x(t_f) - x_f$, where x_f is the desired final state consisting of new desired attitude angles, and zero values for the remainder of the state vector. The function p defines the constraints on the trajectory as well as the control vector (e.g., limits on the elastic motion and the control magnitude). In (21) we have chosen quadratic penalties under the integral sign because they are physically meaningful; they can serve to keep the elastic motion and the rigid-body angular velocity small during the maneuver. The integral part of the performance function (roughly) seeks to minimize the sum of the kinetic energy and the elastic potential energy. However, other functions may be used if appropriate. If the problem is one of minimum-time slewing with control magnitude limit, $u_{\min} \leq u \leq u_{\max}$, and we set $Q = R = S = 0$ in (21), and

$$p_1[x, u, t] = u - u_{\min}$$

$$p_2[x, u, t] = u_{\max} - u$$

Hard constraints in (23) can be incorporated in J by using methods such as Heaviside step weighting [Sag.68]. State and control inequality constraints can yield a bang-bang control law.

Application of Pontryagin's maximum principle [Pon.64] yields the necessary conditions for optimality in the form of a two- point boundary value problem (TPBVP). There is generally no closed- form solution, and numerical methods have to be used to find the optimal $u(t)$ and t_f. Many numerical methods for solving TPBVP are available, and they include: quasilinearization, shooting methods, [Rob.72], method of particular solutions [Mie.70], continuation or homotopy chain methods [Cho.78, Ric.83], and many more. A nice summary of these methods is given in [Jun.86].

This optimal control formulation yields an open-loop control law which is pre-computed under the assumption of perfect model and perfect knowledge of the initial state. Such open-loop control laws can be very sensitive to errors in the knowledge of the plant parameters and the initial state. One method of overcoming the problem would be to superimpose a feedback control law which compensates for deviations from the nominal trajectory [Chu.87]. Alternatively, the problem may be formulated as a feedback control problem by forcing $u(t)$ to be of the form: ("terminal" controller [Jun.86]):

$$u = h[x_f - x(t), t] \tag{24}$$

However, this problem is generally quite difficult to solve.

When the attitude maneuver is relatively small (e.g., on the order of 20°) and if the angular velocities are expected to remain small during the maneuver, it may be appropriate to use a linear LFSS model. In that case, for a fixed t_f, and in the absence of inequality constraint (23), the TPBVP for the minimization problem of (21), (22) can be shown to yield a linear feedback control law with time-varying gain [Jua.86]. A closed form solution can be obtained for that case.

Another way of solving the problem in the linear setting is to remove the terminal manifold constraint in (22). With fixed t_f, the problem reduces to that of designing an optimal finite-duration linear quadratic regulator (LQR). In the linear setting we can take zero as the desired final equilibrium state, and the problem is to transfer state $x(0)$ to the origin. We can raise the terminal penalty S in order to make $x(t_f)$ as small as desired, though it cannot be zero. The advantage of this formulation is that it is computationally much simpler. The control law is given by:

$$u(t) = G(t)x(t) \tag{25}$$

$$G(t) = -R^{-1}B^T P(t) \tag{26}$$

$$\dot{P} = A^T P + PA - PBR^{-1}BP + Q; \quad P(t_f) = S \tag{27}$$

Equation (27) has to be solved backwards from $t = t_f$ to $t = 0$, and $G(t)$ has to be computed and stored for that duration. This approach was used in [Bre.81]. It should be noted, however, that the state vector $x(t)$ cannot be measured for feedback implementation.

An important requirement of a maneuvering control law is that the elastic motion should be held small during the maneuver. One method of accomplishing this is to reduce the high-frequencey components of the input by including a penalty on the first and higher derivatives of the input in the performance function. Denoting $u_0(t) = u(t)$, the state equation is augmented by [Moo.67]

$$\begin{aligned} \dot{u}_0 &= u_1 \\ \dot{u}_1 &= u_2 \\ &\vdots \\ \dot{u}_{k-1} &= u_k \end{aligned}$$

So that the augmented state vector is:

$$x_a = (x^T, u_1^T, u_2^T, ..., u_{k-1}^T)^T$$

A quadratic penalty on x_a will then include quadratic penalties on the first and higher (up to $(k-1)$st) derivatives of $u(t)$, while a quadratic penalty on u_k penalizes the kth derivative of $u(t)$. The effect of this modification is to smooth the optimal control law. The performance function is weighted so as to penalize higher-frequency components of u. A generalization of this concept for other types of frequency weighting is given in [Gup.80].

In principle, feedback control laws are more desirable than open-loop ones. However, the former need accurate knowledge of the complete state vector for feedback. For LFSS, of course, it is not possible to measure the state vector, and state estimators must be used in order to implement the control law. This would result in loss of optimality, even if the LFSS model was accurately known. The presence of modeling errors would increase the state estimation error, possibly to the point of divergence. This would further deteriorate the performance. The issue of non-availability of the state vector has not been adequately addressed in the literature, and remains an open area of research.

A Numerical Example

In order to provide some insight into problems caused by flexibility, we consider a simple single-axis slewing maneuver for the 122m hoop/column antenna. In particular, consider the linear pitch-axis rotational model used in Chapter 3, Sec. 3.3, which consists of the rigid Y-axis rotational mode and the first two Y-axis bending modes. The state vector is: $x = (\theta, \dot{\theta}, q^T, \dot{q}^T)^T$. The objective is to go from the initial attitude angle $\theta(0) = 20^o$ deg. to the final attitude angle $\theta(t_f) = 0$, in 10 seconds. The initial $\dot{\theta}$, q, and $\dot{q}$ are zero, and their desired final values are also zero. The maneuver is to be accomplished using two torque actuators (at locations 1 and 2). The sensor measurements consist of the total (rigid plus elastic) attitude angle and angular rate at the same locations as the actuators. The following cases were considered:

Case a) Bang-Bang Control: Consider the control law given by:

$$\begin{aligned} u(t) &= T_{\max} && 0 \le t \le t_f/2 \\ &= -T_{\max} && t_f/2 < t \le t_f \end{aligned} \tag{28}$$

This is actually the optimal control law for transferring the rigid-body attitude angle form $\theta(0) = \theta_0$ to $\theta(t_f) = 0$ in minimum time, where $|u(t)|$ is constrained to be $\le T_{\max}$. For this example, $T_{\max}$ was calculated to yield $t_f = 10$ sec. for $\theta_0 = 20^o$. The purpose of investigating this control law was to use it as a benchmark for comparing the responses obtained by other methods. Figure 2 shows the inputs, the rigid attitude and rate, and the rotational elastic deflection θ_e and the translational elastic deflection x_e, at location 1. The maximum rotational elastic deflection was about 15° and the maximum translational elastic deflection was about 12 ft., which are excessively large. Therefore, this control law is not acceptable.

Case b) Bang-Bang Control with CDEC: Using differential attitude rate signals at the torquer locations, a CDEC control law (discussed in Chapter 2), was designed to provide a closed-loop damping ratio of 0.5 for the first mode. Because the difference between mode-slopes at the actuator locations was very small for the second mode, the CDEC could not improve its damping. The CDEC control inputs were super-imposed

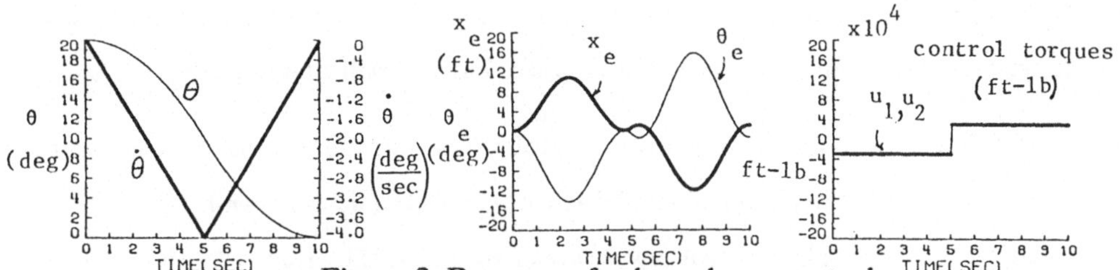

Figure 2. Response for bang-bang control

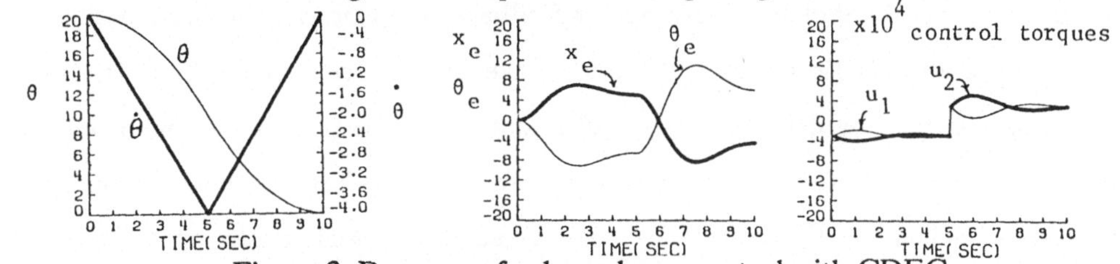

Figure 3. Response for bang-bang control with CDEC

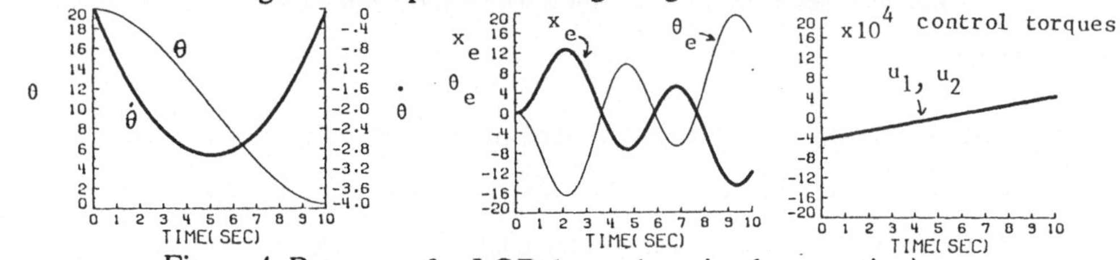

Figure 4. Response for LQR (open-loop implementation)

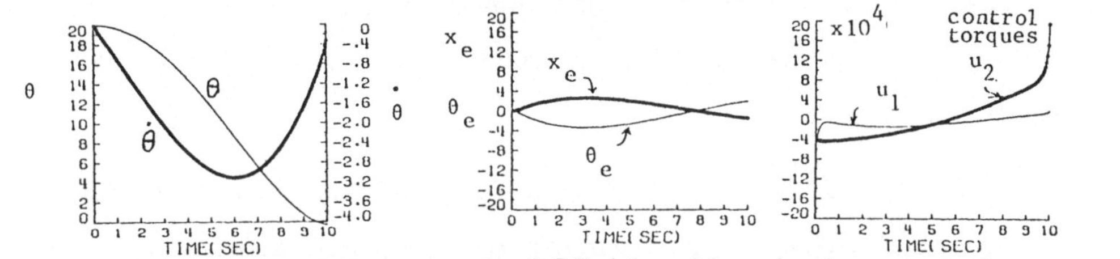

Figure 5. Response for LQR (closed-loop implementation)

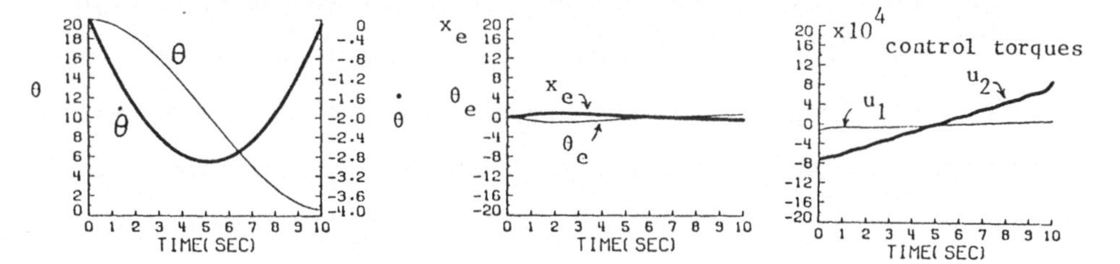

Figure 6. Response for LQR with MESS

on the bang-bang control law of case (a). The resulting responses are shown in Figure 3. The elastic deflections are somewhat smaller, but are still unacceptable.

Case c) LQR Open-Loop Implementation: The finite-duration LQR control law was computed for the "rigid-only" model with $Q = 100I_2$, $R = I_2$, and $\mathrm{S} = \mathrm{diag}(10^{14}, 10^{16})$. The control law is given by (25)-(27). However, the rigid-body attitude and rate cannot be measured for feedback, because the sensors measure the sum of rigid and elastic motion. Therefore, the control law was implemented as an open-loop one, i.e., by using the values of θ and $\dot{\theta}$ predicted via equations of motion. As shown in Figure 4, the elastic deflections were almost as large as those for the bang-bang case. When the CDEC of Case (b) was super-imposed on the open-loop LQR control law, there was about 50% reduction in the elastic deflections, but they were still unacceptable.

Case d) LQR Closed-Loop Implementation: In this case, the LQR was implemented as as a feedback control law, but instead of θ and $\dot{\theta}$, the *measured* attitude and rate (which also include the elastic mode contribution) were used for feeding back. Although the elastic deflections were small (Figure 5), the torque for actuator 2 became very large because of the feedback of elastic motion. Addition of a CDEC resulted only in marginal improvement.

Case e) LQR with MESS: As described in Chapter 3, the model error sensitivity suppression (MESS) method was used to modify the LQR performance function, in order to reduce the control spillover. This was done by adding the term: $\delta_1(\phi_1^T u)^2 + \delta_2(\phi_2^T u)^2$ inside the integral sign in the performance function. Figure 6 shows the responses for $\delta_1 = 10^6$, $\delta_2 = 10^5$. Using the closed-loop implementation, the elastic deflections were reasonably small, and the input torque was well-behaved. Based on these results, the MESS method used with a finite-duration LQR appears to have a good potential for the attitude maneuvering problem.

4.3 Future Research

Improvements are highly desirable in several areas of controller design for LFSS. For fine-pointing controllers, continued research is required for more robust and less conservative designs. The frequency-domain designs of model-based compensators, considered in Chapter 3, were found to be robust with respect to unmodeled elastic mode dynamics, but not with respect to uncertainties in the design model (particularly the natural frequencies). Further research is needed in that area to obtain robust controllers. Design methods which take into account the structure of the uncertainty, (e.g. the structured singular value method [Doy. 84]), offer promise in this regard. Perhaps it would also be possible to capitalize on the special structural properties of LFSS dynamics.

In many cases, a compensator designed for performance and robustness is of high order. It would be desirable to reduce the order of the compensator if it can be done without compromising its performance and robustness. The methods discussed in Chapter 3 for plant order reduction can also be applied to controller order reduction. These methods include: balanced realization [Lau.87], Hankel norm approximation [Glo.84], and stable factorization ([Vid. 87], [Liu.87]). These methods attempt to approximate the full-order compensator $[C(s)]$ with a reduced-order one $[C_r(s)]$, so that the distance: $\| C - C_r \|$ is small. The performance and robustness of the resulting closed-loop system has to be checked after obtaining $C_r(s)$(See [Arm. 88]). An important advantage of the stable factorization method over the other mehtods is that it can be used to approximate stable or unstable systems. In general, it is possible to end up with an unstable model-based compensator; therefore this feature of the stable factorization method is of particular interest. The procedure proposed in [Vid. 87] for order reduction of unstable system, uses normalized right-coprime factorization to obtain a reduced-order model which is close to the original system in the graph topology. When applied to controller order reduction, the method guarantees (under certain conditions) that the resulting reduced-order controller will stabilize the plant provided that the full-order controller stabilizes the plant.

Another approach for controller reduction is to use a formulation which reduces the distance between the closed-loop transfer fuctions rather than between the controllers. That is, $\| G_{c\ell}(s) - G'_{c\ell}(s) \|$ should be minimized, where

$$G_{c\ell} = PC(I + PC)^{-1}, \text{ and } G'_{c\ell} = PC_r(I + PC_r)^{-1} \tag{29}$$

From [And.87],

$$G_{cl} - G'_{cl} \simeq (I + PC)^{-1}[C - C_r](I + PC)^{-1} \tag{30}$$

so that the problem becomes one of minimizing a frequency-weighted H_∞ norm of $[C - C_r]$. Other appropriate frequency weighting may also be used; for example, we may require C_r to approximate C more closely near the frequency ranges corresponding to the bandwidth (low frequency region), as well as where the critical points of the robustness test occur. A nice discussion of many important issues in this topic may be found in [And.87].

Recent developments in H_∞ optimal control theory offer promise for designing compensators for performance and robustness [Fra.87]. The problem can be formulated as that of designing an optimal tracking compensator. For the system shown in Figure 7, the tracking problem can be formulated as an H_∞ minimization problem:

Minimize

$$J = \max_{\|u\|_2 \leq 1} \| W_1 e(t) \|_2 = \| W_1(I + PC)^{-1} W_2 \|_\infty \tag{31}$$

where $W_1(s)$ and $W_2(s)$ are weighting matrices. Suppose (N, D) and $(\tilde{N}, \tilde{D})$ are respectively right and left coprime factorizations of P, i.e., $P = ND^{-1} = \tilde{D}^{-1}\tilde{N}$, where $N, D, \tilde{N}, \tilde{D}$, satisfy the Bezout identities:

$$XN + YD = I \quad \text{and} \quad \tilde{N}\tilde{X} + \tilde{D}\tilde{Y} = I \tag{32}$$

Then the set of all stabilizing compensators is given by:

$$C = (Y - R\tilde{N})^{-1}(X + R\tilde{D}) = (\tilde{X} + DR)(\tilde{Y} - NR)^{-1} \tag{33}$$

where $R(s)$ is a free parameter (any stable proper rational matrix). It can be shown that [Vid.85] the H_∞ minimization problem of (31) can be transformed into that of minimizing

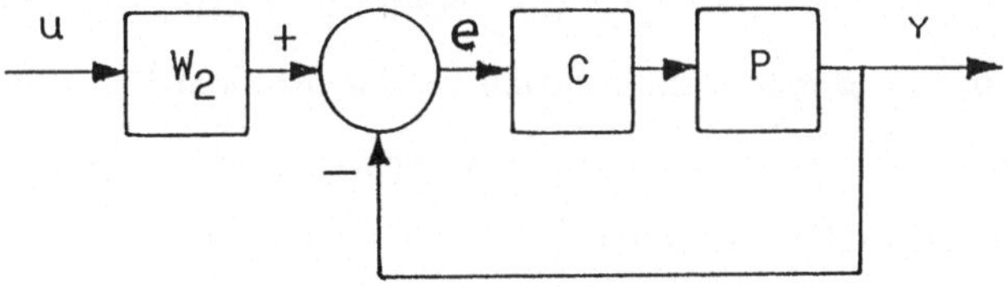

Figure 7. Optimal tracking problem

$$J_\infty = \| F - GRH \|_\infty \tag{34}$$

where

$$F = W\tilde{Y}\tilde{D}W_2, \quad G = W_1 N, \quad H = \tilde{D}W_2 \tag{35}$$

A procedure for solving this problem, which requires inner-outer factorizations of G and H, was given in [Vid. 85]. However, this is a very active area of research, with newer and better results and solution methods coming in at a rapid pace. (for example, see [Doy.88], [Ber.88]).

Future research should investigate the possiblity of using the special structure of LFSS models to obtain closed-form solution to the problem if possible. In fact, it is straightforward to obtain a doubly coprime factorization of the general LFSS model (if the actuators and sensors are collocated):

$$L\ddot{p} + M\dot{p} + Np = \Gamma^T u \tag{36}$$

$$y = \begin{bmatrix} \Gamma & 0 \\ 0 & \Gamma \end{bmatrix} \begin{bmatrix} p \\ \dot{p} \end{bmatrix} = C \begin{bmatrix} p \\ \dot{p} \end{bmatrix} \tag{37}$$

Using the fact that $u = -[I : I]y$ stabilizes the plant [Chapter 3, Sec. 3.1], the results of [Net.84] (along with the basic properties of Laplace transform) can be applied to obtain the following doubly coprime factorization:

$$N = \tilde{N} = CG_0B, \quad D = I - KG_0B, \quad \tilde{D} = I - CG_0F \tag{38}$$

where

$$B = \begin{bmatrix} 0 \\ L^{-1}\Gamma^T \end{bmatrix}, \quad K = [I : I]C \qquad F = \begin{bmatrix} 0 \\ L^{-1}\Gamma^T & L^{-1}\Gamma^T \end{bmatrix} \tag{39}$$

$$G_0(s) = \begin{bmatrix} (Ls^2 + \bar{M}s + \bar{N})^{-1} & (Ls + \bar{M} : L) \\ (Ls^2 + \bar{M}s + \bar{N})^{-1} & [s(Ls + \bar{M}) - I : Ls] \end{bmatrix} \tag{40}$$

$$\bar{M} = M + \Gamma^T\Gamma, \quad \bar{N} = N + \Gamma^T\Gamma \tag{41}$$

Closed-form expressions such as these may parhaps be useful for investigating the possibility of obtaining closed-form solutions to the H_∞ problem.

The design of compensators which are robust to modeling errors constitutes only a part of the overall robustness problem. The final compensator design should also be robust to actuator and sensor nonlinearities, and component failures. Future research should also address the problem of incorporating nonlinearities and failures in the compensator synthesis procedure. It would be highly desirable to investigate the presently available methods for failure detection and isolation, and subsequent controller reconfiguration, for the LFSS control problem. Alternatively, design of fixed compensators which are tolerant to component failures also deserves investigation.

Bibliography

[And.67] Anderson, B. D. O.: A System Theory Criterion for Positive Real Matrices. SIAM J. Control, Vol. 5, 1967.

[And.71] Anderson, B. D. O. and Moore, J. B.: Linear Optimal Control. Englewood Cliffs, N.J.: Prentice-Hall, 1971.

[And.75] Anderson, W. W. and Groom, N. J.: The Annular Momentum Control Device (AMCD) and Potential applications. NASA TN D- 7866, March 1975.

[And.87] Anderson, B. D. O. and Liu, Y.: Controller Reduction: Concepts and Approaches. Proc. 1987 American Control Conference, Minneapolis, MN, June 1987.

[Arb.81] Arbel, A. and Gupta, N. K.: Robust Colocated Control of Large Space Structures. AIAA J. Guidance, Control and Dynamics, Vol. 4, No. 5, Sep.-Oct. 1981.

[Arm.80] Armstrong, E. S.: ORACLS-A Design System for Linear Multivariable Control. Marcel-Dekker, 1980.

[Arm.88] Armstrong, E. S., Joshi, S. M. and Stewart, E. J.: Robust Model-Based Controller Synthesis for the SCOLE Configuration. Proc. 20th Southeastern Symposium on System Theory, Charlotte, N.C., March 1988.

[Ath.66] Athans, M. and Falb, P. L.: Optimal Control, an Inroduction to the Theory and its Applications. McGraw-Hill, New York, 1966.

[Ath.86] Athans, M.: A Tutorial on the LQG/LTR Method. Proc. 1986 American Control Conf., Seattle, WA, June 1986.

[Atl.88] Atluri, S. N. and Amos, A. K. (Eds.): Large Space Structures: Dynamics and Control. Springer-Verlag, NY, 1988.

[Aub.79] Aubrun, J. R., Margulies, G., Arbel, A., and Gupta, N.: Stability Augmentation for Flexible Space Structures. Proc. 18th IEEE Conf. on Decision and Control, Ft. Lauderdale, Florida, Dec. 1979.

[Bai.66] Bailey, F. N.: The Application of Lyapunov's second Method to Interconnected Systems. SIAM J. Contr., Vol. 3, 1966.

[Bal.79] Balas, M. J.: Direct Velocity Feedback Control of Large Space Structures. AIAA J. Guidance & Control, Vol. 2, No. 3, May-Jun. 1979.

[Bal.82] Balas, M. J.: Trends in Large Space Structures Control Theory: Fondest Hopes, Wildest Dreams. IEEE Trans. Auto. Control, Vol. AC-27, No. 3, June 1982.

[Bal.85] Balakrishnan, A. V.: A Mathematical Formulation of the SCOLE Control Problem: Part I. NASA CR 172581, May 1985.

[Bal.86] Balakrishnan, A. V.: Stability Enhancement of Flexible Structures by Nonlinear Boundary Feedback. Proc. IFIP Working Conf. on Boundary Control and Boundary Variations, Nice, France, June 1986.

[Bal.87] Balakrishnan, A. V.: Some Nonlinear Damping Models in Flexible Structures. Fourth SCOLE Workshop, Colorado Springs, Colorado, Nov. 1987.

[Bal.88] Balakrishnan, A. V.: Control of Flexible Flight Structures. Volume dedicated to J. J. Lions. Gauthier-Villari, Paris, 1988.

[Ben.81] Benhabib, R. J., Iwens, R. P., and Jackson, R. L.: Stability of Large Space Structures Control Systems Using Positivity Concepts. AIAA J. Guidance & Control, Vol. 4, No. 5, Sep.-Oct. 1981.

[Ber.88] Bernstein, D.S. and Haddad, W.M.: LQG Control With an H_∞ Performance Bound. Harris Corp. Tech. Rept., Government Aerospace Systems Division, Melbourne, FL, 1988.

[Bre.81] Breakwell, J. A.: Optimal Feedback Slewing of Flexible Spacecraft. J. Guidance & Control, Vol. 4, No. 5, Sep.-Oct. 1981.

[Can.78] Canavin, J. R.: Control Technology for Large Space Structures. Proc. 1980 Joint Automatic Control Conf., San Francisco, Calif., Aug. 1980.

[Cha.84] Chan, S. M. and Athans, M.: Applications of Robustness Theory to Power Systems Models. IEEE Trans. Auto. Control, Vol. AC- 29, No. 1, Jan. 1984.

[Che.61] Chetayev, N. G.: The Stability of Motion (Translated by M. Nadler). Pergamon Press, 1961.

[Cho.78] Chow, S. N., Mallet-Paret, J. and Yorke, J. A.: Finding Zeros of Maps: Homotopy Methods That Are Constructive With Probability One. Math. Comp., Vol. 32, 1978.

[Chu.87] Chun, H. M. and Turner, J. D.: Large Angle Maneuvers for Flexible Spacecraft. Cambridge Research Corp., Final Rept. for NASA Contract NAS1-18098, Oct. 1987.

[Cru.81] Cruz, J. B., Freudenberg, J. S. and Looze, D. P.: A Relationship Between Sensitivity and Stability. IEEE Trans. Auto. Control, Vol. AC-26, No. 1, Feb. 1981.

[Dav.68] Davidon, W. C.: Variance Algorithm for Minimization. The Computer Journal, Vol. 10, No. 4. Feb. 1968.

[Des.75] Desoer, C. A. and Vidyasagar, M.: Feedback Systems: Input- Output Properties. Academic Press, N.Y., 1975.

[Doy.78] Doyle, J. C.: Guranteed Margins for LQG Regulators. IEEE Trans. Auto. Control, Vol. AC-23, No. 4, Aug. 1978.

[Doy.81] Doyle, J. C. and Stein, G.: Multivariable Feedback Design: Concepts for a Classical/Modern Approach. IEEE Trans. Auto. Control, Vol. AC-26, No. 1, Feb. 1981.

[Doy.83] Doyle, J. C. and Wall, J. E.: Characterization of Uncertainty for Large Space Structure Control Problem. Proc. AGARD Conf. on Guidance & Control Techniques for Advanced Space Vehicles, Florence, Italy, Sep. 1983.

[Doy.84] Doyle, J. C.: ONR/Honeywell Workshop on Advances in Multivariable Control (lecture notes). Minneapolis, MN, 1984.

[Doy.88] Doyle, J., Glover, K., Khargonekar, P., and Francis, B.: State Space Solutions to Standard H_2 and H_∞ Control Problems. Proc. 1988 Amer. Control Conf., Atlanta, GA, June 1988.

[Ell.79] Elliott, L. E., Mingori, D. L., and Iwens, R. P.: Performance of Robust Output Feedback Controller for Flexible Spacecraft. Proc. Second VPI & SU/AIAA Symposium on Dynamics and Control of Large Flexible Spacecraft, Blacksburg, Virginia, June 1979.

[Ema.82] Emami-Naeini, E. and Van Dooren, P.: Computation of Zeros of Linear Multivariable Systems. Automatica, Vol. 18, No. 4, July 1982.

[Fra.87] Francis, B. A.: A Course in H_∞ Control Theory. Springer Verlag, N.Y., 1987.

[Gev.70] Gevarter, W. B.: Basic Relations for Control of Flexible Vehicles. AIAA Journal, Vol. 8, No. 4, April 1970.

[Glo.84] Glover, K.: All Optimal Hankel-Norm Approximations for Linear Multivariable Systems and Their L_∞-Error Bounds. International J. Control, Vol. 39, No. 6, 1984.

[Gru.73] Grujic, L. T. and Siljak, D. D.: Asymptotic Stability and Instability of Large Scale Systems. IEEE Trans. Auto. Control, Vol. AC-18, No. 6, Dec. 1973.

[Gup.80] Gupta, N. K.: Frequency-Weighted Cost Functionals: An Extension of Linear Quadratic Gaussian Design Methods. AIAA J. Guiance & Control, Vol. 3, No. 6, Nov.-Dec. 1980.

[Hug.80] Hughes, P. C.: Passive Dissipation of Energy in Large Space Structures. J. Guidance & Control, Vol. 3, No. 4, Jul.-Aug. 1980.

[Hug.81] Hughes, P. C. and Skelton, R. E.: Modal Truncation for Flexible Spacecraft. J. Guidance, Control & Dynamics, Vol. 4, No. 3, May-June 1981.

[Jos.75] Joshi, S. M.: Design of Optimal Partial State Feedback Controllers for Linear Systems in Stochastic Environments. Proc. IEEE Southeastcon, Charlotte, N.C., Apr. 1975.

[Jos.78a] Joshi, S. M.: On Attitude Estmation Schemes for Fine- Pointing Control. IEEE Trans. Aerospace & Electronic Systems, Vol. 14, No. 2, March 1978.

[Jos.79] Joshi, S. M. and Groom, N. J.: Controller Design Approaches for Large Space Structures Using LQG Control Theory. Proc. Second VPI & SU/AIAA Symposium on Dyn. & Contr. of Large Flexible Spacecraft, Blacksburg, Virginia, June 1979.

[Jos.80a] Joshi, S. M. and Groom, N. J.: A Two-Level Controller Design Approach for Large Space Structures. Proc. 1980 Joint Automatic Control Conf., San Francisco, Calif., Aug. 1980.

[Jos.80b] Joshi, S. M. and Groom, N. J.: Optimal Member-Damper Controller Design for Large Space Structures. AIAA J. Guidance & Control, Vol. 3, No. 4, Jul.-Aug. 1980.

[Jos.80c] Joshi, S. M. and Groom, N. J.: Finite Element Structural Model of a Large, Thin, Completely Free, Flat Plate. NASA TM 81887, Sep.1980.

[Jos.81a] Joshi, S. M.: Design of Stable Feedback Controllers for Large Space Structures. Proc. Third VPI & SU/AIAA Symposium on Dyn. and Contr. of Large Space Structures, Blacksburg, Virginia, June 1981.

[Jos.81b] Joshi, S. M.: Control of Large Space Structures Using Annular Momentum Control Devices. Spacecraft Pointing & Position Control: NATO-AGARDO' GRAPH No. 266, Nov. 1981.

[Jos.83] Joshi, S. M.: Control Systems Synthesis for a Large Flexible Space Antenna. Acta Astronautica, Vol. 10, No. 5-6, 1983.

[Jos.84a] Joshi, S. M.: A Modal Model for SCOLE Structural Dynamics. Proc. First Annulal SCOLE Workshop, NASA Langley Research Ctr., Hampton, Virginia, Dec. 1984.

[Jos.84b] Joshi, S. M.: On the Design of Robust LQ Regulators. Proc. 1984 American Control Conf., San Diego, Calif., June 1984.

[Jos.85] Joshi, S. M.: On Robustness Properties of Velocity Feedback Controllers for Large Flexible Space Structures. IEEE Trans. Aerospace & Electronic Sys., Vol. AES-21, No. 1, Jan. 1985.

[Jos.86a] Joshi, S. M.: Robustness Properties of Collocated Controllers for Flexible Spacecraft. AIAA J. Guidance, Control & Dynamics, Vol. 9, No. 1, Jan./Feb. 1986.

[Jos.86b] Joshi, S. M., Armstrong, E. S., and Sundararajan, N.: Application of the LQG/LTR Technique to Robust Controller Synthesis for a Large Flexible Space Antenna. NASA TP-2560, 1986.

[Jos.86c] Joshi, S. M.: An Expression for the Line of Sight Error for the SCOLE Configuration. Proc. Third Annual SCOLE Workshop, Hampton, Virginia, November 1986. (NASA TM-89057, March 1987).

[Jos.87a] Joshi, S. M.: Robustness of Extended-Kalman-Type Observers. International J. Control, Vol. 45, No. 5, 1987.

[Jos.87b] Joshi, S. M.: Design of Failure-Accommodating Multiloop LQG- Type Controllers. IEEE Trans. Auto. Control, Vol. AC-32, No. 8, Aug. 1987.

[Jua.86] Juang, J. N., Turner, J. D. and Chun, H. M.: Closed-Form Recursive Formula for an Optimal Tracker with Terminal Constraints. J. Opt. Theory & Appln., Vol. 51, No. 2, Nov. 1986.

[Jun.86] Junkins, J. L. and Turner, J. D.: Optimal Spacecraft Rotational Maneuvers. Elsevier, N.Y., 1986.

[Kai.80] Kailath, T.: Linear Systems. Prentice-Hall, N.J., 1980.

[Kak.87] Kakad, Y. P.: Dynamics of Spacecraft Control Laboratory Experiment (SCOLE) Slew Maneuvers. NASA CR-4098, 1987.

[Kal.63] Kalman, R. E.: Lyapunov Functions for the Problem of Lur'e in Automatic Control. Proc. Nat. Acad. Sci. (U.S.A.), Vol. 49, Feb. 1963.

[Kap.83] Kappos, E.: Robust Multivariable Control for the F100 Engine. NASA CR-174303, 1983.

[Kel.21] Thomson, W. (Lord Kelvin) and Tait, P. G.: Treatise on Natural Philosophy, Part I. Cambridge University Press, 1921.

[Kie.64] Kiefling, D.: Multiple Vibration Beam Analysis of Saturn I and IB Vehicles. NASA TMX-53072, 1964.

[Kos.83] Kosut, R. L., Salzwedel, H. and Emami-Naeini, A.: Robust Control of Flexible Spacecraft. AIAA J. Guidance, Control & Dynamics, Vol. 6, No. 2, Mar.-Apr., 1983.

[Kwa.72a] Kwakernaak, H. and Sivan, R.: Linear Optimal Control Systems. Wiley-Interscience, New York, 1972.

[Kwa.72b] Kwakernaak, H.: The Maximally Achievable Accuracy of Linear Optimal Regulators and Linear Optimal Filters. IEEE Trans. Auto. Control, Vol. AC-17, Feb. 1972.

[Lau.87] Laub, A. J., Heath, M. T., Paige, C. C. and Ward, R. C.: Computation of System Balancing Transformations and Other Applications of Simultaneous Diagonalization Algorithms. IEEE Trans. Auto. Control, Vol. AC-32, No. 2, Feb. 1987.

[Liu.86] Liu, Y. and Anderson, B. D. O.: Controller Reduction via Stable Factorization and Balancing. International J. Control, Vol. 44, 1986.

[Loz.88] Lozano-Leal, R., and Joshi, S. M.: On the Design of Dissipative LQG-Type Controllers. 27th IEEE Conf. on Decision and Control, Austin, TX, Dec. 1988.

[Mac.77] MacFarlane, A. G. J. and Kouvaritakis, B.: A Design Technique for Linear Multivariable Feedback Systems. International J. Control, Vol. 25, 1977.

[McL.87] McLaren, M. D. and Slater, G. L.: Robust Multivariable Control of Large Space Structures Using Positivity. J. Guidance, Control & Dynamics, Vol. 10, No. 4, Jul.-Aug. 1987.

[Mei.67] Meirovitch, L.: Analytical Methods in Vibrations. McMillan, N.Y., 1967.

[Mei.80] Meirovitch, L.: Computational Methods in Structural Dynamics. Sijthoff and Noordhoff, Rockville, MD, 1980.

[Mei.87] Meirovitch, L. and Quinn, R. D.: Equations of Motion for Maneuvering Flexible Spacecraft. J. Guidance, Control & Dynamics. Vol. 10, No. 5, Sep.-Oct. 1987.

[Mic.72] Michel, A. N. and Porter, D. W.: Stability Analysis of Composite Systems. IEEE Trans. Auto. Control, Vol. AC-17, No. 2, Apr. 1972.

[Mie.70] Miele, A. and Iyer, R. R.: General Technique for Solving Nonlinear Two-Point Boundary Value Problems via the Method of Particular Solutions. J. Optimzation Theory and Applications, Vol. 5, No. 5, 1970.

[Moo.67] Moore, J. B. and Anderson, B. D. O.: Optimal Linear Control Systems With Input Derivative Constraints. Proc. IEEE, Vol. 114, No. 12, Dec. 1967.

[Moo.81] Moore, B. C.: Principal Component Analysis in Linear Systems: Controllability, Observability, and Model Reduction. IEEE Trans. Auto. Control, Vol. AC-26, No. 1, Feb. 1981.

[Net.84] Nett, C. N., Jacobson, C. A. and Balas, M. J.: A Connection Between State-Space and Doubly Coprime Fractional Representations. IEEE Trans. Auto. Control, Vol. AC-29, No. 9, Sep. 1984.

[Nur.84] Nurre, G. S., Ryan, R. S., Scofield, H. N. and Sims, J. L.: Dynamics and Control of Large Space Structures. J. Guidance, Control & Dynamics, Vol. 7, No. 5, Sep.-Oct. 1984.

[Per.82] Pernebo, L. and Silverman, L. M.: Model Reduction via Balanced State-Space Representation. IEEE Trans. Auto. Control, Vol. AC-27, No. 2, April 1982.

[Pin.69] Pinson, L. D. and Leonard, H. W.: Longitudinal Vibration Characteristics of 1/10 Scale Apollo/Saturn V Replica Model. NASA TN-D 5159, April 1969.

[Pon.64] Pontryagin, L. S., Bol'tanski, V. G., Gamkrelidze, R. S. and Mischenko, E. F.: The Mathematical Theory of Optimal Processes. McMillan, N.Y., 1964.

[Pop.73] Popov, V. M.: Hyperstability of Control Systems. Springer- Verlag, N.Y., 1973.

[Ric.83] Richter, S. L. and DeCarlo, R. A.: Continuation Methods: Theory and Applications. IEEE Trans. Circuits & Sys., Vol. 30, No. 6, June 1983.

[Rid.86] Ridgely, D.B. and Banda, S.S.: Introduction to Robust Multivariable Control. AFWAL-TR-85-3102, Feb. 1986.

[Rob.72] Roberts, S. M. and Shipman, J. S.: Two-Point Boundary Value Problems: Shooting Methods. American Elsevier, N.Y., 1972.

[Rob.85] Robertson, D. K.: Three Dimensional Vibration Analysis of a Uniform Beam with Offset Inertial Masses at the Ends. NASA TM 86393, Sep. 1985.

[Ros.74] Rosenbrock, H. H.: Computer-Aided Control System Design. New York Academic, 1974.

[Rus.80] Russell, R. A., Campbell, T. G. and Freeland, R. E.: A Technology Development Program for Large Space Antenna. NASA TM-81902, Sep. 1980.

[Saf.77] Safonov, M. G. and Athans, M.: Gain and Phase Margin for Multiloop LQG Regulators. IEEE Trans. Auto. Control, Vol. AC- 22, No. 2, April 1977.

[Saf.78] Safonov, M. G. and Athans, M.: Robustness and Computational Aspects of Nonlinear Stochastic Estimators and Regulators. IEEE Trans. Auto. Control, Vol. AC-23, No. 4, Aug. 1978.

[Saf.87] Safonov, M. G., Jonckheere, E. A., Verma, M. and Limebeer, D. J. N.: Synthesis of Positive Real Multivariable Feedback Systems. International J. Control, Vol. 45, No. 3, 1987.

[Sag.68] Sage, A. P.: Optimum Systems Control. Prentice-Hall, N.J., 1968

[Sag.71] Sage, A, P. and Melsa, J.: System Identification. Academic Press, N.Y., 1971.

[San.79] Sandell, N. R.: Robust Stability of Linear Dynamic Systems With Application to Singular Perturbation Theory. Automatica, Vol. 15, July 1979.

[Ses.79] Sesak, J. R., Likins, P. W. and Coradetti, T.: Flexible Spacecraft Control by Model Error Sensitivity Suppression (MESS). J. Astronautical Sciences., Vol. 27, No. 2, April-June, 1979.

[Ske.78] Skelton, R. E. and Likins, P. W.: Orthogonal Filters for Model Error Compensation in the Control of Nonrigid Spacecraft. AIAA J. Guidance & Control, Vol. 1, No. 1, Jan.-Feb. 1978.

[Ske.80] Skelton, R. E.: Cost Decomposition of Linear Systems with Application to Model Reduction. International J. Control, Vol. 29, No. 6, 1980.

[Ske.82] Skelton, R. E., Hughes, P. C. and Hablani, H. B.: Order Reduction for Models of Space Structures Using Modal Cost Analysis. J. Guidance, Control & Dynamics, Vol. 5, No.4, Jul.- Aug. 1982

[Sor.80] Sorensen, H. W.: Parameter Estimation: Principles and Problems. Marcel Dekker, N.Y., 1980.

[Ste.87] Stein, G. and Athans, M.: The LQG/LTR Procedure for Multivariable Feedback Control Design. IEEE Trans. Auto. Control, Vol. AC-32, No. 2, Feb. 1987.

[Sul.82] Sullivan, M. R.: LSST (Hoop/Column) Maypole Antenna Development Program, Parts I & II. NASA CR-3558, June 1982.

[Tay.84] Taylor, L. W. and Balakrishnan, A. V.: A Mathematical Problem and a Spacecraft Control Laboratory Experiment (SCOLE) Used to Evaluate Control Laws for Flexible Spacecraft. Proc. First Annual SCOLE Workshop, NASA Langley Research Center, Hampton, Virginia, Dec. 1984.

[Vid.78] Vidyasagar, M.: Nonlinear Systems Analysis. Prentice- Hall, N.J., 1978.

[Vid.80] Vidyasagar, M.: On the Stabilization of Nonlinear Systems Using State Detection. IEEE Trans. Auto. Control, Vol. AC-25, No. 3, June 1980.

[Vid.85] Vidyasagar, M.: Control Systems Synthesis: A Factorization Approach. MIT Press, Cambridge, MA, 1985.

[Vid.87] Vidyasagar, M., Minto, K. D. and Glover, K.: Order reduction of Unstable Systems Using Normalized Coprime Factorization (Draft).

[Wal.67] Walker, J. A. and McClamroch, N. H.: Finite Regions of Attraction for the Problem of Luré. International J. Control, Vol. 6, 1967.

[Wei.68] Weissenberger, S.: Regions of Ultimate Boundedness for the Problem of Luré. Electronics Letters, Vol. 4, No. 18, Sep. 1968.

[Whe.78] Whetstone, W. D.: SPAR Structural Analysis System Reference Manual. NASA CR 158970, Dec. 1978.

[Wyk.66] Wykes, J.H. and Mori, A.S.: An Analysis of Flexible Aircraft Structural Mode Control. Tech. Rept. AFFDL-TR-65-190, Part 1, June 1966.

[You.85] Youseff, A., Wagie, D. A., and Skelton, R. E.: Linear System Approximation via Covariance Equivalent Realizations. J. Math. Analysis & Applcations, Vol. 106, No. 1, 1985.

Index

Lecture Notes in Control and Information Sciences

Edited by M. Thoma and A. Wyner

Vol. 97: I. Lasiecka/R. Triggiani (Eds.)
Control Problems for Systems
Described by Partial Differential Equations
and Applications
Proceedings of the IFIP-WG 7.2
Working Conference
Gainesville, Florida, February 3-6, 1986
VIII, 400 pages, 1987.

Vol. 98: A. Aloneftis
Stochastic Adaptive Control
Results and Simulation
XII, 120 pages, 1987.

Vol. 99: S. P. Bhattacharyya
Robust Stabilization Against
Structured Perturbations
IX, 172 pages, 1987.

Vol. 100: J. P. Zolésio (Editor)
Boundary Control and Boundary Variations
Proceedings of the IFIP WG 7.2 Conference
Nice, France, June 10-13, 1987
IV, 398 pages, 1988.

Vol. 101: P. E. Crouch,
A. J. van der Schaft
Variational and Hamiltonian
Control Systems
IV, 121 pages, 1987.

Vol. 102: F. Kappel, K. Kunisch,
W. Schappacher (Eds.)
Distributed Parameter Systems
Proceedings of the 3rd International Conference
Vorau, Styria, July 6–12, 1986
VII, 343 pages, 1987.

Vol. 103: P. Varaiya, A. B. Kurzhanski (Eds.)
Discrete Event Systems:
Models and Applications
IIASA Conference
Sopron, Hungary, August 3-7, 1987
IX, 282 pages, 1988.

Vol. 104: J. S. Freudenberg/D. P. Looze
Frequency Domain Properties of Scalar
and Multivariable Feedback Systems
VIII, 281 pages, 1988.

Vol. 105: Ch. I. Byrnes/A. Kurzhanski (Eds.)
Modelling and Adaptive Control
Proceedings of the IIASA Conference
Sopron, Hungary, July 1986
V, 379 pages, 1988.

Vol. 106: R. R. Mohler (Editor)
Nonlinear Time Series and
Signal Processing
V, 143 pages. 1988.

Vol. 107: Y. T. Tsay, L.-S. Shieh, St. Barnett
Structural Analysis and Design
of Multivariable Systems
An Algebraic Approach
VIII, 208 pages, 1988.

Vol. 108: K. J. Reinschke
Multivariable Control
A Graph-theoretic Approach
274 pages, 1988.

Vol. 109: M. Vukobratović/R. Stojić
Modern Aircraft Flight Control
VI, 288 pages, 1988.

Vol. 110: In preparation

Vol. 111: A. Bensoussan, J. L. Lions (Eds.)
Analysis and Optimization
of Systems
XIV, 1175 pages, 1988.

Vol. 112: Vojislav Kecman
State-Space Models of Lumped
and Distributed Systems
IX, 280 pages, 1988

Vol. 113: M. Iri, K. Yajima (Eds.)
System Modelling and Optimization
Proceedings of the 13th IFIP Conference
Tokyo, Japan, Aug. 31 – Sept. 4, 1987
IX, 787 pages, 1988.

Vol. 114: A. Bermúdez (Editor)
Control of Partial Differential Equations
Proceedings of the IFIP WG 7.2
Working Conference
Santiago de Compostela, Spain, July 6–9, 1987
IX, 318 pages, 1989

Vol. 115: H.J. Zwart
Geometric Theory for Infinite
Dimensional Systems
VIII, 156 pages, 1989.

Vol. 116: M.D. Mesarovic, Y. Takahara
Abstract Systems Theory
VIII, 439 pages, 1989

Lecture Notes in Control and Information Sciences

Edited by M. Thoma and A. Wyner

Vol. 117: K.J. Hunt
Stochastic Optimal Control Theory
with Application in Self-Tuning Control
X, 308 pages, 1989.

Vol. 118: L. Dai
Singular Control Systems
IX, 332 pages, 1989

Vol. 119: T. Başar, P. Bernhard
Differential Games and Applications
VII, 201 pages, 1989

Vol. 120: L. Trave, A. Titli, A. M. Tarras
Large Scale Systems:
Decentralization, Structure Constraints
and Fixed Modes
XIV, 384 pages, 1989

Vol. 121: A. Blaquière (Editor)
Modeling and Control of Systems
in Engineering, Quantum Mechanics,
Economics and Biosciences
Proceedings of the Bellman Continuum
Workshop 1988, June 13-14, Sophia Antipolis, France
XXVI, 519 pages, 1989

Vol. 122: J. Descusse, M. Fliess, A. Isidori,
D. Leborgne (Eds.)
New Trends in Nonlinear Control Theory
Proceedings of an International
Conference on Nonlinear Systems,
Nantes, France, June 13-17, 1988
VIII, 528 pages, 1989

Vol. 123: C. W. de Silva, A. G. J. MacFarlane
Knowledge-Based Control with
Application to Robots
X, 196 pages, 1989

Vol. 124: A. A. Bahnasawi, M. S. Mahmoud
Control of Partially-Known
Dynamical Systems
XI, 228 pages, 1989

Vol. 125: J. Simon (Ed.)
Control of Boundaries and Stabilization
Proceedings of the IFIP WG 7.2 Conference
Clermont Ferrand, France, June 20-23, 1988
IX, 266 pages, 1989

Vol. 126: N. Christopeit, K. Helmes
M. Kohlmann (Eds.)
Stochastic Differential Systems
Proceedings of the 4th Bad Honnef Conference
June 20-24, 1988
IX, 342 pages, 1989

Vol.127: C. Heij
Deterministic Identification
of Dynamical Systems
VI, 292 pages, 1989

Vol. 128: G. Einarsson, T. Ericson,
I. Ingemarsson, R. Johannesson,
K. Zigangirov, C.-E. Sundberg
Topics in Coding Theory
VII, 176 pages, 1989

Vol. 129: W. A.Porter, S. C. Kak (Eds.)
Advances in Communications and
Signal Processing,
Proceedings of an International Conference
Baton Rouge, Louisiana, October 19-21, 1988
VI, 376 pages, 1989.

Vol. 130: W. A. Porter, S. C. Kak,
J. L. Aravena (Eds.)
Advances in Computing and Control
Proceedings of an International Conference
Baton Rouge, Louisiana, October 19-21, 1988,
VI, 367 pages, 1989

Vol. 131: S. M. Joshi
Control of Large Flexible Space Structures
IX, 196 pages, 1989.